D0058554

THE POLITICS OF THE EARTH

THE POLITICS OF THE EARTH

Environmental Discourses

2ND EDITION

John S. Dryzek

OXFORD
UNIVERSITY PRESS

Great Clarendon Street, Oxford OX2 6DP

Oxford University Press is a department of the University of Oxford.
It furthers the University's objective of excellence in research, scholarship,
and education by publishing worldwide in

Oxford New York

Auckland Cape Town Dar es Salaam Hong Kong Karachi
Kuala Lumpur Madrid Melbourne Mexico City Nairobi
New Delhi Shanghai Taipei Toronto

With offices in

Argentina Austria Brazil Chile Czech Republic France Greece
Guatemala Hungary Italy Japan Poland Portugal Singapore
South Korea Switzerland Thailand Turkey Ukraine Vietnam

Oxford is a registered trade mark of Oxford University Press
in the UK and in certain other countries

Published in the United States
by Oxford University Press Inc., New York

© John S. Dryzek 2005

The moral rights of the author have been asserted

Database right Oxford University Press (maker)

First published 1997

All rights reserved. No part of this publication may be reproduced,
stored in a retrieval system, or transmitted, in any form or by any means,
without the prior permission in writing of Oxford University Press,
or as expressly permitted by law, or under terms agreed with the appropriate
reprographics rights organizations. Enquiries concerning reproduction
outside the scope of the above should be sent to the Rights Department,
Oxford University Press, at the address above

You must not circulate this book in any other binding or cover
and you must impose this same condition on any acquirer

British Library Cataloguing in Publication Data

Data available

Library of Congress Cataloging in Publication Data

Data available

Typeset in Minion and Congress Sans
by RefineCatch Limited, Bungay, Suffolk
Printed in Great Britain
on acid-free paper by
the MPG Books Group

ISBN 978–0–19–927739–1

7 9 10 8 6

PREFACE

A lot has happened in the last four decades of environmental affairs. Environmental crisis arrived in the late 1960s, along with dire warnings about global shortages and ecological collapse. Since then, the Earth's population has increased by over 50 per cent. There have been spectacular nuclear accidents at Three Mile Island and Chernobyl, and spectacular non-nuclear accidents at Bhopal in India and Prince William Sound in Alaska. Green parties have emerged as a significant electoral force, and joined governing coalitions in several countries. Mainstream environmental groups have developed massive memberships. Populist backlashes against environmentalism have flared. Global environmental issues relating to climate change and ozone layer depletion have come to the fore. We have had World Summits, Earth Days, environmental presidents, ecological sabotage, civil disobedience, legislation and regulation by the bookful, and movements for environmental justice, sustainable development, deep ecology, anti-globalization, and 'wise use'.

The idea of this book is to make sense of all these developments. I do so by deploying the notion of environmental discourses. A discourse is a shared way of looking at the world. Its adherents will therefore use a particular kind of language when talking about events, which in turn rests on some common definitions, judgments, assumptions, and contentions. There turns out to be rather little in common between (say) partisans of a discourse believing in the unproblematical nature of uncontrolled economic growth and a radical green discourse seeking renewed harmony among humans and between humans and nature. The history of environmental affairs is largely a matter of the history of the discourses I survey, their rise and fall, their interactions and impacts. As it turns out, all these discourses are still with us, and none has fallen by the wayside (which itself says a lot about the increasing complexity of environmental affairs). I will recount their history, and assess their impact, strengths, and weaknesses as ways of dealing with environmental issues.

I have tried to approach these questions from a position of critical detachment, but at the end of the day I do have some strong positions of my own. I have left an explicit statement of these to the conclusion, under the heading of ecological democracy, though they do put in occasional earlier appearances.

This book began life on September 9, 1994 at 12.45 pm, when Tim Barton of Oxford University Press suggested I write it. I bring this second edition to fruition on September 9, 2004, exactly ten years later. Ruth Anderson of Oxford University Press has helped guide this second edition. In it, I have taken into account and tried to make sense of another eight years of environmental politics and environmental writing, as well as the reactions of readers of the first edition. There are a number of changes in conceptualization—for example, in the discussion of green radicalism in Chapters 9 and 10, where the stress is now on green consciousness and green politics, downplaying an earlier romantic/rationalist continuum that didn't quite work.

The deeper life of this project exists in years of environmental discourse with students, scholars, and activists. In eight years in Oregon I learned much from Joseph Boland, David Carruthers, Irene Diamond, Dan Goldrich, Jeff Land, Gerry Mackie, Michael McGinnis, Ronald Mitchell, Alan Moore, David Schlosberg, Stuart Shulman, Paul Thiers, and Michael Welsh. In Australia, ecopolitical interlocutors have included Mark Carden, Peter Christoff, Steve Dovers, David Downes, Robyn Eckersley, Simon Grant, Carolyn Hendriks, Nicholas Low, Freya Mathews, Simon Niemeyer, Val Plumwood, Adrianna Semmens, Cassandra Star, Richard Sylvan, Janna Thompson, Ken Walker, and David Yencken. Elsewhere, correspondents and conversationalists have included Terence Ball, Brendan Barrett, John Barry, Robert Bartlett, Gary Bryner, Margaret Clark, Tim Clark, Andrew Dobson, Frank Fischer, George González, Robert Goodin, Adolf Gundersen, Garrett Hardin, Bronwyn Hayward, Tim Hayward, Hans-Kristian Hernes, Christian Hunold, Susan Hunter, Michael Jacobs, William Lafferty, Oluf Langhelle, Sang-Hun Lee, James Lester, Stig Toft Madsen, James Meadowcroft, John Meyer, Soon-Hong Moon, Arne Naess, Richard Norgaard, James O'Connor, Claus Offe, Robert Paehlke, Thomas Princen, Craig Rimmerman, Paul Wapner, Albert Weale, Douglas Wilson, Edward Woodhouse, Iris Young, and Oran Young. The particular shape taken by this book depends a lot on advice from Douglas Torgerson and Maarten Hajer (Maarten insists that every book should end with a chapter in which democracy comes to the rescue). I have worked with David Schlosberg on co-editing two editions of the companion reader to this book, *Debating the*

Earth, and benefited a great deal from his insights and advice. Thanks to people such as these, the environmental field is today alive, growing, and the site of some of the most interesting thinking in social science, philosophy, public policy, and practical politics, making a book like this so much easier to write.

In the preface to his classic *Risk Society*, Ulrich Beck says that he wrote most of it overlooking a picturesque lake, and that readers should imagine a lake in the background. I wrote most of the first edition overlooking a garbage dump that is now a park. This second edition was completed in the pleasant surroundings of a dry sclerophyll forest. Readers should imagine tall eucalyptus trees and the call of parrots in the background (but watch out for bush fires).

J.S.D.

Aranda, Australian Capital Territory, September 2004

CONTENTS

PREFACE v

LIST OF BOXES AND FIGURES xiv

PART I INTRODUCTION 1

1 Making Sense of Earth's Politics: A Discourse Approach 3

PART II GLOBAL LIMITS AND THEIR DENIAL 25

2 Looming Tragedy: Survivalism 27

3 Growth Forever: The Promethean Response 51

PART III SOLVING ENVIRONMENTAL PROBLEMS 73

4 Leave it to the Experts: Administrative Rationalism 75

5 Leave it to the People: Democratic Pragmatism 99

6 Leave it to the Market: Economic Rationalism 121

PART IV THE QUEST FOR SUSTAINABILITY 143

7 Environmentally Benign Growth: Sustainable Development 145

8 Industrial Society and Beyond: Ecological Modernization 162

PART V GREEN RADICALISM 181

9 Changing People: Green Consciousness 183

10 Changing Society: Green Politics 203

PART VI CONCLUSION **229**

11 Ecological Democracy **231**

REFERENCES 237

INDEX 255

DETAILED CONTENTS

PREFACE v
LIST OF BOXES AND FIGURES xiv

PART I INTRODUCTION 1

1 Making Sense of Earth's Politics: A Discourse Approach 3

The changing terms of environmental politics 3
A discourse approach 8
Classifying the main environmental discourses 13
Questions to ask about discourses 17
The differences that discourses make 19
The uses of discourse analysis 22

PART II GLOBAL LIMITS AND THEIR DENIAL 25

2 Looming Tragedy: Survivalism 27

The origins of survivalism 27
To the limits—and beyond 30
The political philosophy of survivalism 35
Discourse analysis of survivalism 38
Survivalism in practice 42
Survivalism: an assessment 46

3 Growth Forever: The Promethean Response 51

The Promethean background 51
Promethean argument to the foreground 52
Analysis of Promethean discourse 57
The impact of Promethean discourse 61
Promethean discourse: an assessment 67

PART III SOLVING ENVIRONMENTAL PROBLEMS 73

4 Leave it to the Experts: Administrative Rationalism 75

The repertoire of administrative rationalism 76

Discourse analysis of administrative rationalism 86

The justification of administrative rationalism 89

Administrative rationalism in crisis 92

From government to governance? 96

5 Leave it to the People: Democratic Pragmatism 99

Democratic pragmatism in action 100

Democratic pragmatism as government and governance 108

Discourse analysis of democratic pragmatism 113

The limits of democratic pragmatism 116

6 Leave it to the Market: Economic Rationalism 121

Privatizing everything if you can 123

If you can't privatize it, market it anyway 128

Analysis of economic rationalism discourse 133

An assessment of economic rationalism 137

PART IV THE QUEST FOR SUSTAINABILITY 143

7 Environmentally Benign Growth: Sustainable Development 145

What is sustainable development? 145

The career of the concept 148

Discourse analysis of sustainable development 153

Whither sustainable development? 157

8 Industrial Society and Beyond: Ecological Modernization 162

Cleanest and greenest 162

The idea of ecological modernization 167

Discourse analysis of ecological modernization 169

Radicalizing ecological modernization? 172
Ecological modernization in the balance 176

PART V GREEN RADICALISM **181**

9 Changing People: Green Consciousness **183**
The varieties of green consciousness 183
The romantic disposition and its critics 191
Discourse analysis of green consciousness 193
The impact of green consciousness change 197
Can green consciousness save the earth? 200

10 Changing Society: Green Politics **203**
The varieties of green politics 203
Discourse analysis of green politics 215
Green politics in practice 218
Being green in global capitalist times 225

PART VI CONCLUSION **229**

11 Ecological Democracy **231**

REFERENCES 237
INDEX 255

LIST OF BOXES AND FIGURES

Boxes

1.1	Classifying environmental discourses	15
1.2	Checklist of elements for the analysis of discourses	19
1.3	Checklist of items for assessing the effects of discourses	21
2.1	Discourse analysis of survivalism	41
3.1	Promethean discourse analysis	61
4.1	Discourse analysis of administrative rationalism	89
5.1	Discourse analysis of democratic pragmatism	116
6.1	Discourse analysis of economic rationalism	137
7.1	Discourse analysis of sustainable development	157
8.1	Discourse analysis of ecological modernization	173
9.1	Discourse analysis of green consciousness change	197
10.1	Discourse analysis of green politics	218

Figures

2.1	Exponential growth	32
2.2	Exponential growth with limits	32

PART I

INTRODUCTION

1

Making Sense of Earth's Politics: A Discourse Approach

The changing terms of environmental politics

Over the years, the politics of the Earth has featured a large and growing range of issues. The early concerns were with pollution, wilderness preservation, population growth, and depletion of natural resources. Over time, these concerns have been supplemented by worries about energy supply, animal rights, species extinction, global climate change, depletion of the ozone layer in the upper atmosphere, toxic wastes, the protection of whole ecosystems, environmental justice, food safety, and genetically modified organisms. All these issues are interlaced with a range of moral and aesthetic questions about human livelihood, public attitudes, and our proper relation to other entities on the planet (occasionally even off it). Thus the whole environmental area is home to some heated debates and disputes, ranging from the details of the implementation of policy choices in particular localities, to the arguments of philosophers debating the appropriate ethical position to apply to environmental affairs.

The terms of these debates have changed substantially over time. Consider the following illustrations:

- Once areas of marshy land were called swamps. The only sensible thing to do with swamps was to drain them, so the land could be put to useful purpose. Governments subsidized landowners to drain swamps. Today, we call these same areas wetlands, and governments have enacted legislation to protect their value in providing habitat for wildlife, stabilization of ecosystems, and absorption of pollutants.

- In the nineteenth century, European colonization moved gradually westwards in North America. The United States government provided all kinds of incentives to tame the frontier. Today, the land at the edge of European settlement, which used to be called frontier and was there only to be subdued, is now called wilderness, to be treasured and protected.
- Meanwhile, in Australia and New Zealand, European colonization was followed by the establishment of Acclimatisation Societies to introduce European flora and fauna. These societies approached their task in a spirit of altruism and concern for the public good. Today, governments and citizens in these two countries devote massive effort to the protection of native plants, animals, and ecosystems, and to the extermination of exotic imported species that threaten these ecosystems—imports once cultivated so lovingly by the Acclimatisation Societies.
- After the attacks on the World Trade Center and Pentagon on September 11, 2001, 'terrorists' were confirmed hate figures. Radical environmentalists associated with the Earth Liberation Front were stigmatized as terrorists by the US Federal Bureau of Investigation, and received much longer jail sentences than if they had been classified as 'vandals' in the eyes of the law. Even before 9/11, activist Jeff Leurs was in 2001 sentenced to twenty-two years in prison merely for burning three sports utility vehicles at a dealership in Eugene, Oregon. What exactly is 'eco-terrorism' if the description apples to someone who wants only to destroy ecologically harmful objects, and not hurt people, still less terrorize them?
- What is a whale? Once whales were regarded as sources of food and other useful products such as oil and baleen. The idea that whales were sentient creatures with a right to exist and flourish free from human interference would have been laughable. Yet this view is now widely held, and indeed dominates the policies of most nations on the whaling issue.
- What are people? The idea that there is such a thing as 'population' is a development no more than two hundred years old. Population as an aggregate is something to be controlled and managed: that is, it is more than just 'people.' Given that once there was no such thing as population, the idea of population as a problem, still less population explosion, could not be conceptualized. The Pope, Islamic fundamentalists, and

contemporary anti-environmentalists in the United States still resist conceptualizing population in these terms.

- What is the environment? The environment did not exist as a concept anywhere until the 1960s (though concerns with particular aspects of what we now call the environment, such as open spaces, resource shortages, and pollution do of course pre-date the 1960s). Today, most countries have environmental legislation and government departments with environmental missions, and environmental problems are at the forefront of public attention.
- What is nature? Some radical environmentalists believe that any area modified by human activity is no longer worth caring about. In Edward Abbey's novel *The Monkey Wrench Gang*, one of the environmental heroes measures road distances in terms of six packs of beer, and having finished a can throws it out of the window. The litter is irrelevant, as it ends up in places that have already been destroyed by the construction of a road. Such attitudes horrify more tender-minded environmentalists.
- What, then, is wilderness? One widely held definition is that wilderness consists of land that remains untouched by human extractive activity. But what about the indigenous peoples who have long populated such areas, and in many cases shaped the landscape? And can there be such a thing as wilderness restoration in lands damaged by industrial and agricultural activity?
- What is the Earth? We have long known that it is a planet, but the idea that it might be a finite planet with limited capacities to support human life has only received widespread attention since the late 1960s. Not coincidentally, this was when the Earth was first photographed from space. Since the early 1980s, there has also been a sustained attack on the idea that the Earth is in any sense finite.

The moral of these examples is that contests over meaning are ubiquitous, and the way we think about basic concepts concerning the environment can change quite dramatically over time. The consequences for politics and policies on environmental issues are major. The most basic consequence (to which the last example of the finite Earth points) is that we now have a politics of the Earth, whereas once we did not. If the environment itself were not conceptualized—and it was not, prior to the 1960s—then a book about environmental politics could not be written. Today, of course, we not only have an environment, but most of the important things that happen to it are the subject of politics, and the target of public policy.

Some of the examples I have adduced might seem to suggest that we have a clear trajectory pointing to environmental enlightenment; it is just a matter of humanity becoming more sensitive or aware as time goes on, and escaping from past misconceptions and ignorance. Even if one believes in progress (as I do), it would be a mistake to think of the history of environmental affairs in these terms. What we see instead is that these matters are subject to continuing dispute between people who think in sharply different ways. Some people deny that environmental issues matter at all (how else could President Ronald Reagan have once said that '90 per cent of pollution is caused by trees'?). Consider the following examples of environmental conflicts:

- Citizen-activists in the United States and elsewhere have mobilized in protest against toxic pollutants, responding to seemingly obvious damage to the health of residents and workers. But when scientists employed by government agencies investigate these cases, they typically cannot prove by their own standards that pollution caused death and illness. Activists are rarely persuaded by these results, and continue their campaigns, sometimes winning, sometimes losing. Why is there no consensus on what evidence counts, and what constitutes proof? How should risks be approached in the absence of public confidence in scientific standards?
- The initial growth of the nuclear industry in the 1950s and 1960s took place in secret, away from public concern. By the 1980s, proposals for new nuclear installations were typically the subject of extensive public inquiries, at least in the developed liberal democracies. In Austria, Sweden, and the Netherlands broad national discussions took place in the late 1970s about the whole future of nuclear power, and the kind of society it helped construct. In Britain, inquiries presided over by judges used legalistic rules concerning the admissibility of evidence and argument. Inquiries were focused narrowly on safety issues. It was assumed that the economic benefits of any proposal were positive. Objectors were

not allowed to introduce economic evidence against the proposal, still less arguments about whether nuclear power belongs in a free society, or is consistent with environmental values. The most notorious British nuclear plant is at Windscale/Sellafield on the Irish Sea. A pipeline carries nuclear waste material into the Irish Sea. In 1990 a team of Greenpeace divers placed a symbolic plug in the end of the pipeline. Greenpeace was fined £50,000, and admonished by the judge for being so arrogant as to put their special interests above the law. Why did Britain, in contrast to more progressive European countries, put 'the law' on a pedestal above ecological concerns, rather than trying to integrate ecological principles into the law? Why did 'the law' in Britain consistently serve the interests of the nuclear-industrial complex, and fail to accommodate the kinds of ecological concerns that motivate a group like Greenpeace? Use of the law to suppress ecological activism is not confined to Britain. In April 2002 Greenpeace activists boarded a freighter transporting mahogany cut illegally in Brazilian rainforests to Miami. Stumped for a way to bring Greenpeace to heel, the US Attorney's Office eventually hit upon the idea of charging then with 'sailor-mongering'—a law last used in 1890 against brothel owners who tried to abduct drunken sailors.

- In the United States and Canada the last two decades have seen intense conflict over the logging of remnant old growth forests, especially in the Pacific Northwest. In the United States, logging has been impeded, but by no means halted, by the presence of the spotted owl, an endangered species whose only habitat is the old growth forest. Why is there legislation to protect a species such as the spotted owl (the Endangered Species Act), but no legislation to protect ecosystems such as the forest itself? The conflict between companies and logging communities on the one hand and environmentalists on the other is intense and intractable. Attempts to solve the conflict through the courts, through legislation, and through consensus-seeking exercises (such as the timber summit sponsored by and attended by President Clinton in 1993) have all failed. The George W. Bush administration tried to tip the balance in favor of the timber industry, partly through low-visibility administrative changes, more publicly via the 'Healthy Forests Initiative' passed into law in 2003, that expanded possibilities for logging on public lands, though the stand-off continued. Why is the conflict so intractable? Why do timber workers support logging of old growth to exhaustion instead of sustainable forestry, which would guarantee their jobs and their incomes in the longer term? Can the simultaneous pursuit of environmental and economic values which sustainable forestry connotes actually be achieved? Would this pursuit be secured, as some economists suggest, by dividing the National Forests into chunks of land and selling each chunk to the highest bidder? Why are such proposals, even when their economic logic seems faultless, resisted so strenuously by both environmentalists and loggers?

In all these conflicts, the different sides interpret the issues at hand in very different ways. At any time, the way the issue is dealt with depends largely (though not completely) on the balance of these competing perspectives. In this book I intend making sense of the last forty years or so of environmental concern by mapping these perspectives. But why do these different perspectives exist? Why do debates between their partisans sometimes seem so intractable?

A discourse approach

Environmental issues do not present themselves in well-defined boxes labeled radiation, national parks, pandas, coral reefs, rainforest, heavy metal pollution, and the like. Instead, they are interconnected in all kinds of ways. For example, issues of global climate change due to buildup of carbon dioxide in the atmosphere from burning fossil fuels relate to air pollution in more local contexts, and so to issues of transportation policy. These issues also relate to destruction of the ecosystems (such as tropical forests) which act as carbon sinks, absorbing carbon dioxide from the atmosphere; and to issues of fossil fuel reliance and exhaustion; and so to problems related to alternative sources of energy such as nuclear power. Thus environmental problems tend to be interconnected and multidimensional; they are, in a word, complex. Complexity refers to the number and variety of elements and interactions in the environment of a decision system. When human decision systems (be they individuals or collective bodies such as governments) confront environmental problems, they are confronted with two orders of complexity. Ecosystems are complex, and our knowledge of them is limited, as the biological scientists who study them are the first to admit. Human social systems are complex too, which is why there is so much work for the ever-growing number of social scientists who study them. Environmental problems by definition are found at the intersection of ecosystems and human social systems, and thus are doubly complex.

The more complex a situation, the larger is the number of plausible perspectives upon it—because the harder it is to prove any one of them wrong in simple terms. Thus the proliferation of perspectives on environmental problems that has accompanied the development and diversification of environmental concern since the 1960s should come as no surprise. It is my intention here to make sense of this proliferation. I shall do so by deploying the notion of 'discourse.'

A discourse is a shared way of apprehending the world. Embedded in language, it enables those who subscribe to it to interpret bits of information and put them together into coherent stories or accounts. Discourses construct meanings and relationships, helping to define common sense and legitimate knowledge. Each discourse rests on assumptions, judgments, and contentions that provide the basic terms for analysis, debates, agreements, and disagreements. If such shared terms did not exist, it would be hard to imagine problem-solving in this area at all, as we would have continually to return to first principles. The way a discourse views the world is not always easily comprehended by those who subscribe to other discourses. However, as I will show, complete discontinuity across discourses is rare, such that interchange across discourse boundaries can occur, however difficult.

Discourses are bound up with political power. Sometimes it is a sign of power that actors can get the discourse to which they subscribe accepted by others. Discourses can themselves embody power in the way they condition the perceptions and values of those subject to them, such that some interests are advanced, others suppressed (Foucault, 1980). Discourses are also intertwined with some material political realities. Governments in capitalist economies have to perform a number of basic functions whether they want to or not (see Dryzek, 1992*a*): first and foremost, ensuring continued economic growth. Corporations can stop investing in response to government policies they do not like. The increasing mobility of capital and finance intensifies this pressure, because businesses can threaten to transfer their operations to countries with less stringent environmental policies and practices. Just south of the United States–Mexico border is a zone of *maquiladora* industries, producing for US markets without having to worry about US anti-pollution laws, still less about Mexican laws that look good on paper but are never enforced. Thus the first task of governments is to keep actual and potential corporate investors happy. If governments make investors unhappy—through (say) tough anti-pollution policy—then they are punished by disinvestment, which in turn means recession, unpopularity in the eyes of voters, and falling tax revenues. Often the reason investors take such actions is that they subscribe to a particular discourse that defines some government policies as right, others as wrong.

Now, trying to make sense of the Earth's politics through reference to discourses is not the only way of going about the task. Other analysts look at the institutions (markets, government bureaucracies, legal systems, etc.) that have been developed for handling environmental issues.[1] Some look at the policies that governments have pursued. Some care little about the details of real-world practices, focusing instead on the political philosophies that can be applied in environmental affairs. Some look only at particular case studies of environmental issues. I shall have plenty to say about institutions, policies, political philosophies, and case studies, for all owe much to the discourses in their vicinity.

This inquiry rests on the contention that language matters, that the way we construct, interpret, discuss, and analyze environmental problems has all kinds of consequences. My intent is to lay out the basic structure of the discourses that have dominated recent environmental politics, and present their history, conflicts, and transformations. I intend to produce something more than just an account of environmentalism. Environmental discourse is broader than that, extending to those who do not consider themselves environmentalists, but either choose or find themselves in positions where they are handling environmental issues, be it as politicians, bureaucrats, corporate executives, lawyers, journalists, or citizens. Environmental discourse even extends to those who consider themselves hostile to environmentalism. My geographic coverage for the most part encompasses Europe, North America, Australasia, and the global arena; though sometimes it is appropriate to look elsewhere.

Some studies in this idiom examine discourse carefully in the context of a particular issue. That such an approach is productive is demonstrated in studies by Maarten Hajer of transformations in discourse on acid rain in Britain and the Netherlands in the late 1980s and early 1990s (Hajer, 1995), and of a recent discourse of 'nature development' in Dutch environmental policy (Hajer, 2003). Equally productively, Karen Litfin has elucidated changing international discourse about global ozone layer depletion in the 1980s (Litfin, 1994). However, there is room for breadth as well as depth in analyzing environmental discourse, for looking at the big picture rather than the details. My own accounts will lack the richness of Hajer and Litfin inasmuch as I cannot always say exactly who said what and why behind which closed doors to whom about a particular point, and how the other responded.

In offering a view of a much bigger territory, I will be guided by some analytical devices and distinctions (to be introduced shortly) that give me some confidence in painting such large and complex discursive terrain in broad strokes. I seek vindication only in the plausibility of the stories I tell. These stories are backed by my own twenty-five years of working and teaching in the environmental field, but others might still carve up the territory somewhat differently. For example, Andrew Dobson (1990) makes a threefold distinction between old-fashioned conservationism, reform environmentalism, and radical ecologism. Robyn Eckersley (1992) thinks that the key difference is between anthropocentric (human-centered) and ecocentric perspectives. Relatedly, in histories of US environmentalism, it is standard practice to distinguish between two traditions, heirs respectively to the anthropocentric rational resource management advocated by the US Forest Service's first chief forester Gifford Pinchot and the deeper respect for nature propounded by Sierra Club founder John Muir (see, for example, Taylor, 1992). To Martin Lewis (1992), the only distinction that makes sense is between moderates and extremists, or 'Promethean' and 'Arcadian' environmentalists as he styles them (I will use the term Promethean somewhat differently). Less worthy of serious attention, former US Secretary of the Interior James Watt distinguished between environmentalists and Americans.

Discourse is important, and conditions the way we define, interpret, and address environmental affairs. This should not be taken to mean that there is only discourse when it comes to environmental problems.

Postmodernists believe that there is no escape from specific viewpoints (for an environmental application, see Bennett and Chaloupka, 1993), such that 'nature' and 'wilderness' are mainly social constructions, understood culturally as the product of societies that have, among other things, removed indigenous peoples from their landscapes. But even those such as Cronon (1995) and Soper (1995) who make this argument also stress that their position does not diminish environmental concern. Thus nature should not be treated as merely a subcategory of culture, as an extreme postmodern position would require. Such an extreme position would be just another anthropocentric turn in the colonization of nature for human purposes (Crist, 2004), an arrogance that fails to recognize nature's existence prior to human appropriation.

Just because something is socially interpreted does not mean it is unreal. Pollution does cause illness, species do become extinct, ecosystems cannot absorb stress indefinitely, tropical forests are disappearing. But people can make very different things of these phenomena and—especially—their interconnections, providing grist for political dispute. The existence of these competing understandings is why we have environmental politics (or any kind of politics) to begin with. Sometimes particular constructions can be exposed as misguided—as, for example, when automobile company executives in the 1950s dismissed the possibility of smog in cities such as Los Angeles by claiming that car exhaust emissions were simply absorbed by the atmosphere. More often, it is hard to prove constructions right or wrong in any straightforward way. But one might say the same about scientific worldviews, political ideologies, or governmental constitutions. It is still possible to engage in critical comparative judgment, to apply evidence and argument, and to hope that in so doing we can correct some errors, and so move toward a better overall understanding of environmental issues and problems. As Litfin puts it, it is possible to subscribe to both a hermeneutic epistemology (i.e., an interpretive philosophy of inquiry) and a realist ontology (i.e., a commitment to the actual existence of problems) (1994: 26–7, 50).[2] In analysis and argument, appeals to what is 'natural' are often made, but there is no single uninterpreted 'nature' capable of putting an end to political dispute.

Unfortunately, there are plenty of forces that can impair critical comparative judgment. The public relations departments of large corporations are adept here when it comes to 'greenwashing' their activities. According to *O'Dwyers PR Services*, a major public relations journal, the environment is 'the life and death PR battle of the 1990s' (*Guardian*, London, September 18, 1996), and this is no less true in the new millennium. In the 1990s, the Weyerhauser Corporation advertised itself as 'the tree growing company'™, before changing its 'tagline' in 1999 to 'the future is growing'™. Weyerhauser does plant and grow a lot of trees. But the trees it plants are single-species plantations, managed with herbicides and pesticides. Many of the trees it cuts down are in multi-species old growth forests, which take hundreds of years to mature. Corporate front groups often have names that connote environmental concern. But the real intent of the Global Climate Coalition, for example, is to put a spin on climate issues that is conducive to the short-term interests of oil companies (see Rowell, 1996). Similar stories apply to the National Wetland Coalition, Alliance for Environment and Resources, and National Wilderness Institute (see Ehrlich and Ehrlich, 1996: 11–23).

Alternatively, such actors can sponsor discourses of environmental concern conducive to their own interests. Perhaps this is why so many of them find the idea of sustainable development and its potential commitment to continued economic growth so attractive (as we will see in Chapter 7).

Classifying the main environmental discourses

Environmental discourse begins in industrial society, and so has to be positioned in the context of the long-dominant discourse of industrial society, which we can call industrialism. Industrialism may be characterized in terms of its overarching commitment to growth in the quantity of goods and services produced and to the material wellbeing that growth brings. Industrial societies have of course featured many competing ideologies, such as liberalism, conservatism, socialism, Marxism, and fascism. But whatever their differences, all these ideologies are committed to industrialism. From an environmental perspective they can all look like variations on this theme. This commonality might surprise their adherents, more conscious of their ideological differences than of their industrialist commonalities. But all these ideologies long ignored or suppressed environmental concern. If what we now call environmental issues were thought about at all, it was generally in terms of inputs to industrial processes. For example, rational use of such inputs was the main concern of the Conservation Movement founded at the beginning of the twentieth century in the United States, whose key figure was Gifford Pinchot. This movement did not want to preserve the environment for aesthetic reasons, or for the sake of human health. Instead, the Conservation Movement sought only to ensure that resources such as minerals, timber, and fish were used wisely and not squandered, so that there would always be plenty of them to support a growing economy.

Environmental discourse cannot therefore simply take the terms of industrialism as given, but must depart from these terms. This departure can be reformist or it can be radical; and this distinction forms one dimension for categorizing environmental discourses.

A second dimension would take note of the fact that departures from industrialism can be either prosaic or imaginative. Prosaic departures take the political-economic chessboard set by industrial society as pretty much given. On that chessboard, environmental problems are seen mainly in terms of troubles encountered by the established industrial political economy. They require action, but they do not point to a new kind of society. The action in question can be quite dramatic and radical. As we will see, there are those who believe that economic growth must be reined in, if not brought to a halt entirely, in order to respond effectively to environmental problems. But the measures endorsed or proposed by these people are essentially those which have been defined by and in industrialism. For example, those who would curb economic growth normally propose that this be done by strong central administration informed by scientific expertise—a quintessentially industrialist instrument.

In contrast, imaginative departures seek to redefine the chessboard. Notably, environmental problems are seen as opportunities rather than troubles. Imaginative redefinition of the chessboard may dissolve old dilemmas, treating environmental concerns not in opposition to economic ones, but potentially in harmony. The environment is brought into the heart of society and its cultural, moral, and economic systems, rather than being seen as a source of difficulties standing outside these systems. The thinking is imaginative, but the degree of change sought can be small and reformist, or large and radical. As we shall see, imaginative reformist ways of rendering the basic political-economic structure bequeathed by industrial society capable of coping with environmental issues may be found.

On the other hand, imaginative radical changes can also be envisaged, requiring wholesale transformation of this political-economic structure. Combining these two dimensions—reformist versus radical and prosaic versus imaginative—produces four cells, as indicated in Box 1.1.

Environmental problem solving is defined by taking the political-economic status quo as given but in need of adjustment to cope with environmental problems, especially via public policy. Such adjustment might take the form of extension of the pragmatic problem-solving capacities of liberal democratic governments by facilitating a variety of environmentalist inputs to them; or of markets, by putting price tags on environmental harms and benefits; or of the administrative state, by institutionalizing environmental concern and expertise in its operating procedures. Within the overall discourse of environmental problem solving there may be substantial disagreement as to which of these forms is appropriate. So, for example, a debate between proponents of administrative regulation and market-type incentive mechanisms for pollution control has been under way since the 1970s, and shows few signs of letting up.

Survivalism is the discourse popularized in the early 1970s by the efforts of the Club of Rome (which I will discuss in the next chapter) and others, still retaining many believers. The basic idea is that continued economic and population growth will eventually hit limits set by the Earth's stock of natural resources and the capacity of its ecosystems to support human agricultural and industrial activity. The limits discourse is radical because it seeks a wholesale redistribution of power within the industrial political economy, and a wholesale reorientation away from perpetual economic growth. It is prosaic because it can see solutions only in terms of the options set by industrialism, notably, greater control of existing systems by administrators, scientists, and other responsible elites.

BOX 1.1 Classifying environmental discourses

	Reformist	Radical
Prosaic	Problem solving	Survivalism
Imaginative	Sustainability	Green radicalism

Sustainability commences in the 1980s, and is defined by imaginative attempts to dissolve the conflicts between environmental and economic values that energize the discourses of problem solving and limits. The concepts of growth and development are redefined in ways which render obsolete the simple projections of the limits discourse. There is still no consensus on the exact meaning of sustainability; but sustainability is the axis around which discussion occurs, and limits lose their force. Without the imagery of apocalypse that defines the limits discourse, there is no inbuilt radicalism to the discourse. The era of sustainability begins in earnest with the publication of the Brundtland Report in 1987 (World Commission on Environment and Development, 1987). At the same time, ideas about ecological modernization, seeing economic growth and environmental protection as essentially complementary, arose in Europe.

Green radicalism is both radical and imaginative. Its adherents reject the basic structure of industrial society and the way the environment is conceptualized therein in favor of a variety of quite different alternative interpretations of humans, their society, and their place in the world. Given its radicalism and imagination, it is not surprising that gre-n radicalism features deep intramural divisions—to which I shall attend. In the United States, social ecologists with a pastoral vision and a concern for social justice debate deep ecologists, who prefer landscapes without humans. In Germany, Green *Fundis* eventually lost a struggle with Green *Realos* over tactical questions about action in the streets versus action in parliament. Everywhere, green romantics disagree with green rationalists, proponents of the rights of individual creatures disagree with more holistic thinkers, and advocates of green lifestyles disagree with those who prefer to stress green politics. These debates are lively and persistent; but the disputants have far more in common with each other in terms of basic dispositions, assumptions, and capabilities than they do with either industrialism or with the three competing discourses of environmental concern just introduced.

These, then, are the four basic environmental discourses, and I will organize the chapters that follow according to how they fit with these four categories. All four reject industrialism; but all four engage with the discourse of industrialism—if only to distance themselves from it. And this is why their engagement with industrialism and its defenders is often more pronounced than their engagement with each other.

Questions to ask about discourses

So far I have identified the four basic discourses in fairly general terms. But in order to see why and how these discourses have developed, and to what effect, it is necessary to pin down their content more precisely. This I shall do in the chapters that follow. To this end, let me now develop a set of questions for the analysis of discourses.

Discourses enable stories to be told; in fact, the title of a discourse can be an abbreviated storyline (the concept of environmental storylines is deployed by Hajer, 1995). To refer back to the four discourses just enumerated, limits or survivalism connotes a story about the need to curb ever-growing human demands on the life-support capacities of natural systems. Problem solving connotes a different story—indeed, can subsume a number of different stories—about the unpleasant side-effects of particular economic activities requiring piecemeal remedies. Each discourse constructs stories from the following elements.

Basic entities whose existence is recognized or constructed

This is what is meant by the 'ontology' of a discourse. Different discourses see different things in the world. Some discourses recognize the existence of ecosystems, others have no concept of natural systems at all, seeing nature only in terms of brute matter. At least one other entertains the idea that the global ecosystem is a self-correcting entity with something like intelligence. This is the idea of Gaia, which I will address in my analysis of green radicalism. Some discourses organize their analyses around rational, egoistic human beings; others deal with a variety of human motivations; others still recognize human beings only in their aggregates such as states and populations. Most believe it is fruitful to deal with 'humans' as a category, a few that it is necessary to break down on the basis of gender. Some assume governments and their actions matter; others believe it is the human spirit that is crucial.

Assumptions about natural relationships

All discourses embody notions of what is natural in the relationships between different entities. Some see competition, be it between human beings in markets or between creatures locked in Darwinian struggle, as natural. Others see cooperation as the essence of both human social systems and natural systems. Hierarchies based on gender, expertise, political power, species, ecological sensibility, intellect, legal status, race, and wealth are variously assumed in different discourses; as are their corresponding equalities.

Agents and their motives

Story lines require actors, or agents. These actors can be individuals or collectivities. They are mostly human, but can be nonhuman. In one discourse we may find benign and public-spirited expert administrators. Another discourse might portray the same people as selfish bureaucrats. Still others might ignore the presence of government officials altogether. Many other kinds of agents and motives put in appearances. They include enlightened elites, rational consumers, ignorant and short-sighted populations, virtuous ordinary citizens, a Gaia that may be tough and forgiving or fragile and punishing, among others.

Key metaphors and other rhetorical devices

Most storylines, in the environmental arena no less than elsewhere, depend crucially on metaphor. Key metaphors that have figured in environmental discourse include:

- spaceships (the idea of 'spaceship earth');
- the grazing commons of a medieval village ('the tragedy of the commons');
- machines (nature is like a machine that can be reassembled to better meet human needs);
- organisms (nature is a complex organism that grows and develops);
- human intelligence (ascribed to nonhuman entities such as ecosystems);
- war (against nature);
- goddesses (treating nature in benign female form, and not just as Mother Nature).

Metaphors are rhetorical devices, deployed to convince listeners or readers by putting a situation in a particular light. Many other devices can perform the same tasks. These include appeal to widely accepted practices or institutions, such as established rights, freedoms, constitutions, and cultural traditions. For example, the rights of species, animals, or natural objects can be justified through reference to the long-established array of individual human rights in liberal societies. Appeals can be made to deeper pasts, such as pastoral or even primeval idylls, as a way to criticize the industrial present. The negative and discredited can be accentuated as well as the positive and treasured. For example, it is possible to collect horror stories about government mistakes on environmental issues, and sprinkle these horror stories into arguments. On the other hand, some discourses collect and accentuate success stories.

This completes my checklist of items for the scrutiny and analysis of discourses. The items are summarized in Box 1.2. If my discussion of each element has been brief, matters should become clearer when I deploy this checklist in subsequent chapters in order to capture the various discourses more precisely. Beyond capturing the essence of the various discourses and their subdivisions, it is of course important to determine what difference each of them makes. I have already asserted that the language we use in addressing environmental affairs does make a difference, but this needs to be demonstrated for particular discourses, rather than just asserted as a general point.

The differences that discourses make

With this need to demonstrate the implications of different discourses in mind, I will take a look at the history as well as the content of each

BOX 1.2 Checklist of elements for the analysis of discourses

1. Basic entities recognized or constructed
2. Assumptions about natural relationships
3. Agents and their motives
4. Key metaphors and other rhetorical devices

discourse. This history can generally be traced back to some aspect of industrialism—if only as a rejection of that aspect. With time, environmental discourses develop, crystallize, bifurcate, and dissolve. A crucial part of this history consists of the kind of politics surrounding, shaping, and shaped by the discourse. In some cases the politics might be that of a social movement or political party; in other cases, governmental commissions and intergovernmental negotiations; in others, administrative control; in others, elite bargaining; in others, rationalistic policy design. Sometimes there will be little in the way of politics at all, as, for example, in the case of 'lifestyle' greens. Sometimes the politics may be local, sometimes national, sometimes transnational, sometimes global.

The impact of a discourse can often be felt in the policies of governments or intergovernmental bodies, and in institutional structure. For example, the flurry of environmental legislation enacted in many industrialized countries around 1970 mostly reflected a discourse of administrative rationalism (a sub-category of what I have defined as problem solving). Since 1970, problem-solving discourse has also been embodied in a number of institutional innovations that extend the openness and reach of liberal democratic control of environmental affairs (in the form of devices such as public inquiries and various procedures for consensual dispute resolution). Beyond affecting institutions, discourses can become embodied in institutions. When this happens, discourses constitute the informal understandings that provide the context for social interaction, on a par with formal institutional rules. Or to put it slightly differently, discourses can constitute institutional software while formal rules constitute institutional hardware. Sometimes, though, discourses do not have direct effects on the policies or institutions of governments, but take effect elsewhere. For example, green radicalism has helped some individuals and communities to distance themselves from both government and corporate capitalism in putative attempts to create an alternative political economy relying on self-sufficiency. Impacts can also be felt directly on society and culture without having to pass through formal institutions or public policies. Contemporary social movements often target the way ordinary people think and behave, and much of their success can be judged in these terms. For example, feminism has changed the division of labor in households. Environmentalism has led many people to change their lifestyles so as to reduce their ecological impact, be it through recycling, buying organic food, avoiding genetically modified organisms, planting native vegetation around their houses, using public transport rather than private cars, or boycotting companies with a poor environmental record.

To assess more fully the worth and impact of a discourse requires attention to its critics as well as its adherents. Sometimes, adherents of different discourses will ignore and dismiss rather than engage one another. Nevertheless, dispute does occur across the boundaries of different discourses. Frequently, this occurs between the environmental discourse in question and the older discourse of industrialism. Given that each of the four categories of discourse I have identified has its roots in either modification or conscious rejection of industrialism, this is not too surprising. Occasionally, debate is engaged between the problem-solving, limits, sustainability, and green radical discourses. If such engagement is infrequent, that is mostly a matter of these four discourses viewing issues and problems in such different ways that little interchange across their boundaries can occur. One goal of this book is to promote such interchange.

Attention to the arguments of critics will facilitate identification of flaws in the discourse. Such identification will also be helped by attention to experience of the practical implications of the discourse, in politics, policies, institutions, and beyond. The tools of discourse analysis which I have enumerated enable further critical analysis of the promise and peril attached to each discourse in its contribution to environmental debate, analysis, and action. It may even turn out that there are some complementarities between different discourses, rather than simple rivalry.

The set of questions I will ask in order to assess the impact, plausibility, and attractiveness of each discourse is summarized in Box 1.3.

BOX 1.3 Checklist of items for assessing the effects of discourses

1. Politics associated with the discourse
2. Effect on policies of governments
3. Effect on institutions
4. Social and cultural impact
5. Arguments of critics
6. Flaws revealed by evidence and argument

The uses of discourse analysis

As should be clear, my intent is to advance analysis in environmental affairs by promoting critical comparative scrutiny of competing discourses of environmental concern. This intent distances me from some others who have developed and deployed discourse analysis.

The concept of discourse in the sense I am using it owes much to the efforts of Michel Foucault (for example, 1980), who revealed the content and history of discourses about illness, sex, madness, criminality, government, and so forth. Foucauldians are generally committed to the idea that individuals are for the most part subject to the discourses in which they move, and so are seldom able to step back and make comparative assessments and choices across different discourses. It should be evident that I disagree. Discourses are powerful, but they are not impenetrable (as Foucault and his readers have themselves demonstrated in their own exposé of the history of various discourses). Foucault and his followers also often portray discourses in hegemonic terms, meaning that one single discourse is typically dominant in any time and place, conditioning not just agreement but also the terms of dispute. Along these lines, Luke (1999) treats environmentalism mostly in terms of an 'environmentality' that actually serves rather than disrupts the established order of industrial society. In contrast, I believe that the variety found in environmental discourses is important. The environmental arena reveals that the discourse of industrialism was indeed hegemonic, to the extent that 'the environment' was hardly conceptualized prior to the 1960s. However, this hegemony eventually began to disintegrate, yielding the range of environmental discourses now observable. While in its totality environmentalism can be positioned as a challenge to industrialism, it does not constitute a unified counter discourse. Rather, environmentalism is composed of a variety of discourses, sometimes complementing one another, but often competing. With the necessary preliminaries over, it is to a mapping of these discourses and their consequences that I now turn.

NOTES

1 For my own contribution to this genre, see Dryzek, 1987.

2 The position I take here is consistent with a critical realist philosophy of science (Bhaskar, 1975), for which real structures exist, while our understanding of them is limited by selective inquiry, exposure, and experience.

PART II

GLOBAL LIMITS AND THEIR DENIAL

Environmental issues can be as local as the dog droppings on the grass in front of my house, or as global as the greenhouse effect. When environmental issues made their first dramatic leap to the top of the political agenda in the late 1960s, it was the global issue that really captured public attention. Not coincidentally, this was also the first time the Earth was photographed from space, and a beautiful, fragile place it looked. For the first time in human history the Earth could be conceptualized readily as a finite planet, and for the first time a true politics of planet Earth became conceivable. Environmental problems were soon cast in terms of threats to the capacity of this planet to support life—especially human life.

The threats in question involved degradation of the global environment through pollution, and exhaustion of the Earth's natural resources (fossil fuels, minerals, fisheries, forests, and croplands). Urgency came from population explosion and economic growth. Exponential growth in both human numbers and the level of economic activity meant that there was no time to lose, for humanity seemed to be heading for the limits at an ever-increasing pace. Hitting these limits would mean global disaster and a crash in human populations.

This discourse of limits and survival was given a major boost by the Club of Rome, an international organization composed of industrialists, politicians, and academics. The Club's most famous product was a set of computer-generated projections of the global future published in 1972 in the international best seller *The Limits to Growth*. These projections showed in graphical terms that if humanity continued on its profligate course then it had at the very most a century before disaster would strike. There were many calls for radical action to stop this headlong rush to destruction, though the survivalists' political repertoire turned out to consist mainly of some tried-and-tested practices, especially strong governmental control.

Survivalism met with immediate counterattack from defenders of the established industrial economy, whose taken-for-granted order of things survivalism had challenged. These defenders argued that humans are characterized by unlimited ingenuity, symbolized in Greek mythology by the progress made possible by the theft of fire from the gods by Prometheus. Prometheans asserted that the Earth was in truth unlimited; that as soon as one resource threatened to run out, ingenious people would develop a substitute. This had always happened in the past, and it would continue to happen in the future. The Promethean reaction gathered speed in the 1980s, for it fit quite well with the ideological climate of the Reagan years in the United States. The administration of President George W. Bush beginning in 2001 was still more open to Promethean assumptions.

The dispute between the two camps continues, and neither shows any sign of conceding. Yet it matters crucially which side is right. If the Prometheans are correct, then not only is survivalism wrong, but environmentalism of any kind simply loses its urgency. So: who is right?

2

Looming Tragedy: Survivalism

The origins of survivalism

Population biologists and ecologists have long deployed the concept of 'carrying capacity'—the maximum population of a species that an ecosystem can support in perpetuity. According to the population biologist and survivalist Garrett Hardin, the ecologist's Eleventh Commandment is 'Though shalt not transgress the carrying capacity' (1993: 207). When the population of a species grows to the point where carrying capacity is exceeded, the ecosystem is degraded and population crashes, recovering only if and when natural processes restore the ecosystem to its previous capacity. Such crashes are readily observable in relatively simple ecosystems, for example when large herbivores such as deer are introduced to environments with no predators. Their populations soon explode to the point where the food supply is exhausted.

When population biologists turn to human affairs they see identical possibilities (see especially Catton, 1980). Historical examples of human transgression of carrying capacity are readily observable in the denuded landscapes of the Mediterranean littoral. The concept of carrying capacity has to be adjusted downward inasmuch as quality rather than quantity of human lives becomes an issue—an ecosystem can support more humans at subsistence level than it can with any greater quality of human life. The waters are muddied still further when it comes to contemporary human populations, because trade and aid mean that human societies can escape constraints imposed by the carrying capacities of geographically bounded ecosystems. For example, biologists show that the carrying capacity of East African ecosystems for human beings has been exceeded; but continued food aid from overseas means that population does not crash. This sort of analysis is controversial, as it calls into question some basic humanitarian (population biologists would say soft-headed) impulses on the part of donors, not to mention some radical analyses of international political economy; but those critiques can wait.

The claim that local environmental scarcities in the developing world can lead to violence and misery as ethnic groups, social classes, or nations struggle over dwindling resources is central to the idea of 'environmental security' (Homer-Dixon, 1999). However, the fact that civil conflicts such as those endemic to Haiti and Chiapas in Mexico occur in degraded environments does not prove that environmental degradation is itself a cause of these conflicts, and even proponents of the environmental security thesis are careful to say scarcity is just one factor in conflicts (Homer-Dixon, 1999: 7).

Given trade and aid, it really makes most sense to talk in terms of the human carrying capacity of the global ecosystem. Analyses at regional, national, or local levels soon get into all kinds of conceptual angles. It really matters little that Singapore, New York City, London, and Los Angeles have vastly exceeded the carrying capacity of their local ecosystems, so long as they can exploit distant resources and sinks for their pollutants in order to support large and sometimes growing human populations. This is not to excuse such cities for their vast 'ecological footprints.'

The other complicating factor when it comes to applying population biology to human societies is the possibility of economic growth. Unique among animals, the ecological burdens imposed by each human are not roughly constant—indeed, they can seem virtually unbounded. Think, for example, of the conspicuous consumption of the super-rich such as Michael Jackson, Donald Trump, or Bill Gates. So if the number of humans is growing and the amount consumed per human is growing, the ecological news is not good. Now, there are those who argue that economic growth is good for the environment because it allows some of the financial fruits of growth to be diverted to conservation; but such arguments belong in a different discourse, and will be dealt with in the next chapter. Let me stick for the moment to the discourse of ecological limits.

This sort of analysis, which anticipates misery, starvation, and death resulting from unconstrained human procreation and consumption, and that sees the main political challenge as ensuring some level of human survival at an adequate level of amenity, is not new. It goes back to William Forster Lloyd (1794–1852) and, more famously, Thomas Malthus (1766–1834), widely reviled as the 'dismal parson.' Free-market liberal economists dismissed him because he doubted cherished Victorian notions of material progress and wealth accumulation leading to general social improvement. Marxists and other socialists were more scornful still, as he cast doubt on their postulate of material plenty in a free and equal post-capitalist society, where individuals define their own needs and the means for their satisfaction. Ultimately, Malthus's argument was falsified by two hundred years of population growth combined not with misery, but with rising living standards (at least in what is now the developed world). Or so it would seem.

In 1968 Garrett Hardin published his enormously influential essay, 'The Tragedy of the Commons.' Hardin's analysis quickly became part of the toolkit for analysts of environmental problems. His analytics had in fact been a staple of resource economics for some time (see Gordon, 1954). Hardin himself paid homage to his more distant precursor, William Forster Lloyd.[1] Unlike the economists, Hardin had the good sense to give the analysis a catchy name, publish in the large-circulation journal *Science*, refrain from graphs and algebra, and put it out just as the widespread perception of environmental crisis hit for the first time.

Hardin's logic of the commons is straightforward. Facing a decision about whether or not to put an extra cow on the village commons, each rational self-interested peasant will recognize that the benefits of the extra cow accrue to himself alone, whereas the costs (stress upon the commons) are shared with the other villagers. Thus all villagers will quickly put more cows on the commons, which will in turn be destroyed. Hardin was using the commons of a medieval village as a metaphor for all kinds of environmental resources (the process he described never actually happened in any medieval village, to our knowledge). So each decision maker deciding whether or not to catch an additional netful of fish, or dump an additional ton of sewage, or cut down a tree, or drive an extra mile in Los Angeles, or get that malfunctioning catalytic converter fixed, is facing essentially the same decision: private benefit and the public interest point in opposite directions. Hardin made a connection to childbearing decisions: if the world is a commons, each additional child adds stress to the commons, even though calculations of private interest determine that the child should be conceived, born, and raised.

Of course, all this is only tragic if the commons is finite—that is, if there are limits. If there are no limits, we can populate, grow, and consume at will. For several centuries it had seemed that unconstrained economic growth was the natural order of things, and social survival in finite systems was simply not conceptualized. But come 1970, everything changed in a hurry. The world looked as though it was being hit by what Paul Ehrlich (1968) sensationalized as a population bomb, more powerful than nuclear bombs. Combined with economic growth, population explosion was going to exhaust stocks of energy, cropland, clean water, minerals, and the assimilative capacity of the atmosphere and oceans. Matters were dramatized further by the energy crisis which arrived suddenly in 1973 with the oil embargo organized by the Organization for Petroleum Exporting Countries, to pressure the industrial world to take an anti-Israel line in the Middle East conflict.

To the limits—and beyond

This discourse of limits and survival was not all there was to environmentalism circa 1970. Many concerns were more local, more aesthetic, more concerned with quality of life than its mere perpetuation. But survivalism did set the apocalyptic horizon of environmentalism, the basic reason why concern about the environment was not just desirable, but also necessary. This is no less true after 2000 than in the late 1960s. In the early 1970s, though, survivalism did more than set these horizons. One of the most heated debates in the history of environmental concern was occasioned in 1972 by the publication of *The Limits to Growth* (Meadows et al., 1972), which within four years had sold four million copies.

This famous study was sponsored by the Club of Rome, founded in 1968 by well-heeled industrialists and sympathetic academics concerned with the 'predicament of mankind.'[2] Curiously, these industrialists were keen to show that industrialism itself might be unsustainable. The Massachusetts Institute of Technology team commissioned by the Club of Rome to undertake the study were not population biologists who knew about carrying capacity, or economists who might know a bit about economic growth, but systems modelers (with business school backgrounds). Systems dynamics was a new technology rendered useful by the development and availability of computers. The MIT team was in a sense 'Malthus with a computer' (Freeman, 1973), though unlike Malthus they believed the predicament of mankind had more causes than just population growth. The exercise was legitimated by computers—for both analysis and salvation. As Torgerson (1995: 9) notes, the film version of *The Limits to Growth* portrays a dramatic contrast between the chaos and misery of a world seemingly hell-bent on using resources to exhaustion, polluting till it choked, with population exploding into misery, on the one hand, and the calm authority symbolized by a computer and the experts who could run it on the other. Those days were of course well before the appearance of user-friendly personal computers and anarchistic hackers; computers were large, slow, and complicated.

The computer runs themselves were simulations over a hundred years or more into the future of predicted pathways of key aggregates, which interacted with one another through a host of interrelated variables. The key aggregates were resources, population, industrial output, food supply, and pollution. (Critics quickly pointed out the absence of technology and prices.) All variables were measured at the global level. The predictions varied somewhat depending on the assumptions built into different computer runs, but given postulated limits to resource availability, agricultural productivity, and the capacity of the ecosphere to assimilate pollution, some limit was generally hit within a hundred years, leading to the collapse of industrial society and its population. The policy prescription was obvious: humanity needed to change its profligate ways if it was to avoid the apocalypse of overshoot and collapse. Meadows and colleagues envisaged an alternative 'stationary state' global economy, at a fixed level of throughput of resources, with a stable population. As they note, this state had been envisaged more than a hundred years earlier by the political economist John Stuart Mill (p. 175). Within these constraints, economic growth is not ruled out. Quite how humanity could move to this steady state was less clear, though as we shall see others quickly supplied the political prescriptions.

The elaborate computer simulations really stated the obvious: exponential growth cannot go on forever in a finite system. Exponential growth is growth at a constant percentage rate. This produces a growth curve over time of the sort depicted in Figure 2.1. With this curve, it matters little whether a limit such as the global supply of natural resources

FIGURE 2.1 **Exponential Growth**

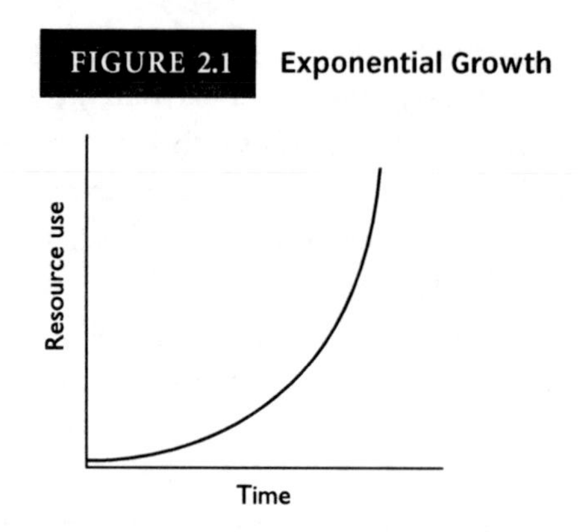

is doubled, tripled, or quadrupled; a little more time may be bought, but the limit is hit soon enough (see Figure 2.2). Nor should we expect much advance warning, for under exponential growth the limit will (by definition) be approached at an absolute speed unprecedented in human history.

Lester Brown (1978) deploys the metaphor of *The Twenty-Ninth Day* to ask on what day a pond will be half-covered with lilies if the coverage doubles every day, and will cover the whole pond on the thirtieth day. The answer is, of course, the twenty-ninth day. It would be very easy to look at the pond on the twenty-ninth day and conclude there is plenty of clear water. The challenge is to figure out how to build a capacity for foresight into collective decision making, before any evidence of global collapse is apparent. Currently, decision making in government and business is almost entirely geared to the short term.

FIGURE 2.2 **Exponential Growth with Limits**

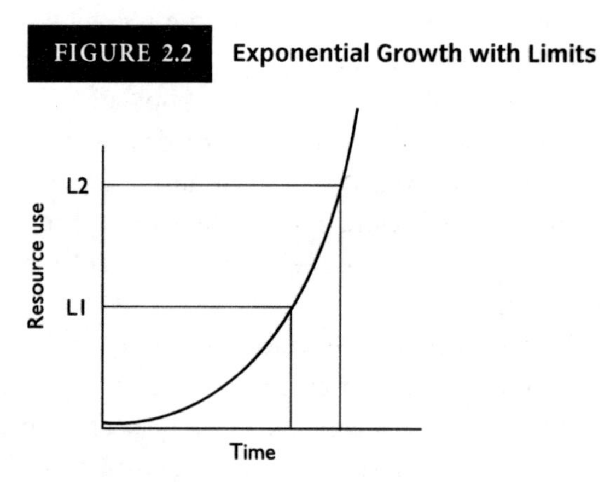

Note: Doubling the available supply of resources from L1 to L2 buys very little time.

Global modeling continued under the auspices of the Club of Rome, the United Nations, and the United States government, among others. Under President Jimmy Carter, the US government's various global modeling enterprises were integrated in the *Global 2000 Report to the President*, a gloomy report for a gloomy presidency. As its title implies, *Global 2000* did not look as far ahead as *The Limits to Growth*, but only to the year 2000. Its major findings are summarized in the opening paragraph of Volume 1:

> If present trends continue, the world in 2000 will be more crowded, more polluted, less stable ecologically, and more vulnerable to disruption than the world we live in now. Serious stresses involving population, resources, and environment are clearly visible ahead. Despite greater material output, the world's people will be poorer in many ways than they are today.

With exquisitely bad timing, *Global 2000* was released in 1980, just in time for the arrival in Washington DC of Ronald Reagan, with an entourage and a worldview that could not fathom such pessimism. Still, the Reagan presidency and the exuberant era it symbolized did not completely silence the survivalist discourse, even in the United States, even in its capital.

In Washington, focus on the parlous state of key global aggregates was kept alive by the Worldwatch Institute, under the leadership of Lester Brown. With its annual *State of the World* reports, beginning in 1984, the Institute keeps reminding us that indicators of environmental quality and resource availability point in the wrong direction, and that disaster is just around every corner. Though its view of the world is more nuanced than that of the *Limits to Growth*, the emphasis remains on monitoring systems and aggregates at the global level. The main systems are forests, grasslands, fisheries, and croplands. The overall limit identified by the Worldwatchers is the photosynthetic energy these systems can make available for human use. Humans currently appropriate a large and growing proportion of this energy, wasting much of it. The challenge is to use photosynthetic energy more efficiently.

In 1992, twenty years of survivalism were celebrated by an updated *Limits to Growth* entitled *Beyond the Limits* (Meadows et al., 1992), whose major theme was that little had changed except the passage of twenty wasted years. In 2002 the US National Academy of Sciences published a report concluding that total human demands on the biosphere exceeded carrying capacity (Wackernagel et al., 2002). Authors such as Paul Ehrlich,

Lester Brown, Norman Myers, and Garrett Hardin continued to contribute to the survivalist discourse in the 1990s and 2000s (Brown, 2003; Ehrlich and Ehrlich, 2004). In 1995 a few leading economists broke ranks with their colleagues (once near-unanimous in their condemnation of limits) to announce that economic growth sooner or later must encounter the environment's carrying capacity (see Arrow et al., 1995). These ecological economists endorsed the basic survivalist tenet that the Earth's resource base is finite, and called for institutional redesign to reduce stress on natural systems.

The field of ecological economics should not be confused with an older environmental economics. Environmental economics is the handmaiden of economic rationalism, discussed at length in Chapter 6. Ecological economics, in contrast, treats the environment not as an adjunct to the human economy, nor as a mere medium through which some human actions harm or benefit other humans. Instead, ecosystems are conceptualized as the fundamental entities within which human economic systems are embedded. Thus, environmental problems are to be thought of as shortfalls in the capacity of interdependent ecological systems and economic systems acting in conjunction to sustain human—and possibly non-human—life. Ecological economics treats natural systems as finite, and so the scale of human economic activity which they can support becomes an issue. The main challenge is to figure out how economic systems could be sustainable within these constraints.

While meeting resistance (or indifference) within the mainstream discipline of economics, ecological economics is gradually acquiring academic respectability, through the International Society for Ecological Economics and its journal *Ecological Economics*, both founded in 1989.

The pioneers of ecological economics include Nicholas Georgescu-Roegen (1971) and Herman Daly (1977). Georgescu-Roegen explored the implications of the second law of thermodynamics, which specifies that any closed system will over time deteriorate in the direction of sameness or disorder without external input of energy. The fact that there is only a limited supply of low entropy or order on this planet has major economic implications. Low entropy is really the ultimate form of scarcity. It exists in mineral structures, concentrated fossil fuels, in ecosystems; but human economic activity is running down the supply of low entropy.

Herman Daly worked out principles of steady state economics.

Conventional economics is committed to perpetual economic growth, and indeed sees economic health and normality in terms of the presence of growth. Daly rose to the challenge presented by the survivalist denial of unlimited growth, describing how an economy could be run on steady-state lines, without requiring ever-increasing environmental and natural resource inputs (for an accessible treatment of the steady state, see also Czech, 2000).

The basic idea of global limits is shared by many if not most green radicals. Their critic Martin Lewis (1992: 9–10) credits survivalism with being the mainspring of green radicalism—as he puts it: 'green extremism is rooted in a single, powerful conviction that continued economic growth is absolutely impossible, given the limits of a finite planet.' However, the politics of green radicalism prove to be very different from survivalism.

The political philosophy of survivalism

Many of the more prominent figures who have furthered the discourse of limits and survival have a background in biology. Their severe policy prescriptions can obliquely commit their authors to particular political structures. For example, Ehrlich (1968) countenances compulsory sterilization in countries such as India, which could hardly be effective without a more authoritarian politics than India featured then or now.[3] Missing in many analyses and associated public debates is sustained attention to the politics, economics, and—crucially—political economy of survival. Where such analysis should be, one often finds wishful thinking. For example, in *Beyond the Limits* Meadows et al. (1992: 222–36) conclude with prescriptions for visioning, networking, truth-telling, learning, and loving. Quite how these all might play out in the real-world political economy is less clear. In fairness to Meadows and colleagues they do address the deficiencies of the market, noting that it is 'blind in the long term and pays no attention to ultimate sources and sinks until it is too late' (p. 184; see also, much earlier, Meadows, 1976).

Other authors have produced much more in the way of detailed, bold, and striking political-economic analyses. Especially in the 1970s, and in some cases continuing to the present, survivalist political prescriptions have been centralized and authoritarian. Garrett Hardin stands out among the limits biologists in his willingness to tackle questions of political-economic organization very explicitly. If, as he avers in his classic essay, 'freedom in the commons brings ruin to all' (1968), then obviously freedom, including the freedom to breed, needs to be curtailed. The solution to the tragedy of the commons in situations characterized by limits is 'mutual coercion mutually agreed upon,' whether the commons is a local fishery or the global atmosphere. Hardin (1977) expresses skepticism about effective central authority above the level of the nation-state. Taking survivalism quite literally, he recommends that the more developed countries abandon the underdeveloped world if governments in the latter wish to continue policies that promote population explosion and ecological devastation. Developed nations would then constitute 'lifeboats' afloat in a world otherwise drowning in misery. Hardin's argument here downplays the fact that it is the developed nations, not the poor ones, that impose the greater stresses on the world's ecosystems. Survivalism could equally support a program of wealth reduction in rich countries to support redistribution to poor countries within global ecological limits (Reuveny, 2002: 84).

Other survivalists are less obviously callous than Hardin, though in basic agreement with him that abuse of the commons, resource exhaustion, and environmental despoliation are largely a matter of individuals and other actors pursuing material interests in decentralized systems. Decentralized systems have no cohesive leadership directing them: examples include markets, liberal democratic political systems, and the international system. In such systems there is no incentive to care about collective goods like environmental quality or long-term human well-being. Thus Robert Heilbroner in 1974 concluded that the only hope for humanity lies in monastic government combining 'religious orientation with a military discipline' (Heilbroner, 1991: 176–7; first edn. 1974) in order to cure humanity's profligate ways. Such totalitarian government would control economic transactions as well as politics. Obviously, authoritarian governments committed to industrialism rather than environmental conservation are no help, as the disastrous environmental records of dictatorships around the world testify.

The need to control access to the commons is just one prop for authoritarian government in the limits discourse. A second prop comes with the fact that the discourse places great emphasis on expertise, of systems modellers, population biologists, and ecologists. In some cases, the relevant knowledge is quite hard to master: ecology does, of course, deal in very complex systems. This sort of recognition leads William Ophuls, in the most comprehensive and sophisticated analysis of the political ramifications of ecological crisis to appear in the 1970s, to recommend establishment of a governing class of 'ecological mandarins' (1977: 163). The expertise is 'on top' rather than 'on tap,' because ecology's claim to primacy means that there is no room for tradeoff against competing values of the sort that ordinary politicians routinely seek. Ophuls also echoes Hardin in that management of the commons needs strong central authority.[4]

Later survivalists would soften this authoritarianism. For example, ecological economists mainly seek only intelligent and creative use of some well-understood policy instruments. Michael Jacobs (1991) details how sustainability planning can work through government creation and enforcement of environmental quality standards. These standards might refer to air quality, biodiversity, or the wellbeing of the ozone layer. Using models of the impact of economic activities on these indicators, governments could enact policies that limit the damage. Daly (1992) is quite keen on the tradeable pollution permit approach developed by economic rationalists (see Chapter 6), on the grounds that governments can specify the overall amount of pollution allowed within, say, a watershed based on ecological principles, but then let the market operate to determine who should pollute how much within this overall limit.

Some survivalists have even become positively enthusiastic about democracy and citizen action. So Richard Barnet (1980) catalogued limits, but proposed democratic mobilization, not authoritarianism, to confront them. Norman Myers, carrying the limits torch prominently and ably into the 1990s, issued a clarion call for citizen action to confront limits (in his contribution to Myers and Simon, 1994). Lester Brown has consistently placed greater faith in localized citizen action than in national leadership (Brown, 1981; Brown et al., 1992: 180), though this faith sits uneasily with a plea for a stronger United Nations to 'do for all people what national governments cannot' (Brown et al., 1992: 179). Czech (2000: 114) hopes in the face of all evidence to the contrary that: 'In the world's model democracy [the United States], the steady state revolution must be a revolution in public opinion, a process by which the virtually ubiquitous cherishing of economic growth is transformed into an equally ubiquitous castigation of economic growth.' Such calls to citizen action are far indeed from the oligarchy proposed by survivalist political theorists in the 1970s. These theorists did, though, remain unrepentant in the 1990s (see Hardin, 1993; Heilbroner, 1991; Ophuls and Boyan, 1992).

To see why broad-based citizen action can only fit uneasily into the discourse of limits and survival, a closer examination is in order, using the tools of discourse analysis introduced in Chapter 1.

Discourse analysis of survivalism

The basic storyline of survivalism is that human demands on the carrying capacity of ecosystems threaten to explode out of control, and draconian action needs to be taken in order to curb these demands. This storyline is in turn constructed from the following components.

Basic entities whose existence is recognized or constructed

Survivalism recognizes and emphasizes the resources upon which human beings depend for their existence. These include stocks of non-renewable resources, such as oil, gas, coal, metallic ores, and cropland. Ecosystems are recognized in the basic survivalist ontology, but only in limited fashion: as sources of renewable resources such as firewood, timber, soil, and fish, or as sinks for the absorption of pollution. Crucially, stocks of non-renewable resources and the capacity of ecosystems to produce renewable resources and assimilate wastes are treated as finite. The discourse also emphasizes human population as an aggregate entity (i.e., it is something other than just 'people'), whose size and growth has all kinds of implications for human destiny. Finally, elites—especially those associated with governments, and especially those with pertinent expertise, be it in systems modeling, ecology, or population biology—play a central role.

Finite stocks of resources, ecosystems as founts of renewable resources and sinks for pollutants, population, and elites—all these might sound unremarkable, just what one would expect to find in environmental talk.

In fact, these items together constitute a highly selective and, as I will show later in this chapter, deeply problematical set. For the moment, though, it should be noted that this basic ontology is not exhaustive (ontologies never are). Missing, for example, are individual problem solvers, human beings as social creatures capable of devising cooperative arrangements, markets (except to be dismissed), social movements, genders, resilient ecosystems, states, and interest groups.

Assumptions about natural relationships

The relationships assumed by survivalists to be the most natural in human affairs are conflict and hierarchy. Conflict ranges from rivalry in access to the commons to struggle over scarce resources. Often conflict and hierarchy are taken for granted. In contrast, the survivalists I have discussed who analyze alternative forms of political-economic organization, such as Hardin, Heilbroner, and Ophuls, reason their way toward hierarchy rather than take it for granted. The basis for hierarchy can be expertise, or virtue, or both. Certainly, human beings conceptualized en masse as 'population' do not have the required virtue to control their appetites or their procreation. Garrett Hardin argues at length that conscience is self-eliminating, for those without a conscience will have more children (1968). (This argument relies on the controversial premise that desired number of children is a hereditary trait.) Survivalists such as Lester Brown and Norman Myers later softened this hierarchical commitment to include pleas for widespread citizen action. These pleas notwithstanding, the survivalist discourse deals ultimately in aggregates such as population, resource stocks, global pollution levels, and, crucially, monitoring and control of these aggregates. Such control is hard to envisage on anything other than a hierarchical basis. The hierarchy does not need to be the tight, authoritarian style of government proposed by some survivalists; but it does require coordinated action on the part of elite groups.

Agents and their motives

It is, then, elites who have agency, the capacity to act. Their motivations are up for grabs. Elites can choose to operate national political economies according to established principles of maximizing economic growth, leavened by a touch of social justice and the need to placate special interests; or they can choose to oversee the transition to a stationary state through coordinated global action. 'Populations,' be they national, global, or class-specific, have no agency; they are only acted upon, as aggregates to be monitored through statistics and controlled by government policy. At most, their component individuals can only follow their short-sighted desires (though they are allowed to be quite rational in this blinkered sense).

Key metaphors and other rhetorical devices

Survivalism is rich in metaphors. These include, first and foremost, the notion of overshoot and collapse, drawn from models of simple ecosystems where one species breeds to excess and then experiences a crash. The tragedy of the commons is rooted in metaphor: Garrett Hardin made his original argument in the context not of any resource currently threatened with exhaustion, but rather the common land of a medieval village. Another favorite metaphor is the spaceship, introduced by Kenneth Boulding (1966). If the life-support systems of the spaceship are not maintained, the crew dies. 'Spaceship earth' became a credible notion as real spaceships left the earth with humans aboard for the first time and—crucially—these humans photographed the earth from space. This image gave powerful impetus to thinking about the earth as a whole system—and a finite, fragile one at that. A photograph of the earth from space graces the cover of many books on my 'environmental' shelf (but not this one).

In their account of biodiversity, Paul and Anne Ehrlich (1981) deploy a metaphor of a plane held together by rivets. If one rivet falls out (one species becomes extinct) nothing happens; but as more rivets pop out, the plane eventually crashes. Other metaphors capture the nature of exponential growth: the pond whose surface covered by lilies doubles every day, the population bomb, population explosion. Czech (2000) compares resource exploitation to 'shoveling fuel for a runaway train' destined for a crash. Cancer as a metaphor has occasionally been invoked. So Peccei refers to 'the cancerous growth of population' (1981: 29; see also Gregg, 1955). The obvious inference is that we human beings are the cancer cells upon the body of the earth. While this is perhaps not a metaphor calculated to have broad appeal beyond (tiny) VHEMT, the Voluntary Human Extinction Movement, Paul Ehrlich did once draw the further inference that 'the cancer itself must be cut out' (Ehrlich, 1968: xi). In a 2003 speech, UK Environment Minister Michael Meacher likened the human race to a virus that could destroy the Earth.[5] Davidson (2000) wants to moderate survivalism by replacing all these metaphors with one that likens environmental degradation to pulling threads from a tapestry. Rents appear, the quality suffers, but the tapestry never actually falls apart.

A key rhetorical device in the 1970s was the computer. Ostensibly, the computer was used to carry out complex calculations about the interaction of a host of variables; in fact, these computer models did little more than state the obvious, that exponential growth cannot proceed indefinitely in a finite environment. Later, computers became far less mysterious, and the computer lost its rhetorical power in survivalist discourse. If the computer represented the rationalistic, calculating side of survivalist discourse, it coexists uneasily with quasi-religious images of doom and redemption. The earthly paradise of a stationary state is attainable—but only if we recognize our sin, and change our ways.

BOX 2.1 **Discourse analysis of survivalism**

1. **Basic entities recognized or constructed**
   - Finite stocks of resources
   - Carrying capacity of ecosystems
   - Population
   - Elites
2. **Assumptions about natural relationships**
   - Conflict
   - Hierarchy and control
3. **Agents and their motives**
   - Elites; motivation is up for grabs
4. **Key metaphors and other rhetorical devices**
   - Overshoot and collapse
   - Commons
   - Spaceship Earth
   - Lily pond
   - Cancer
   - Virus
   - Computers
   - Images of doom and redemption

Survivalism in practice

What difference has survivalism made in environmental affairs? And has any such influence been for the better or for the worse? Survivalism provides the apocalyptic horizon of environmental concern, raising the stakes in environmental affairs. In these terms, its effects may be profound, while hard to trace directly into particular politics, policies, or outcomes. Certainly the bleak authoritarian prescriptions proposed by the survivalist vanguard (especially in the 1970s) find little reflection in political practice or institutional design, and show few signs of being adopted anywhere. The draconian population control attempted briefly and disastrously in India in the mid-1970s, and in China with greater effect since then, may constitute exceptions. The more pervasive political practice associated with survivalism turns out to take a form somewhat different from the authoritarian political theory.

This form does, though, remain an elitist affair, certainly not that of a social movement, or even interest groups involving large numbers of people. Survivalism treats most people as 'population,' effectively denying them agency, the capacity to act. The politics of limits is exclusive. Now, pressure groups devoted to furthering the survivalist agenda have existed and do exist, in the form of organizations such as Negative Population Growth (NPG), the Club of Rome (which later turned to ecological modernization; see Chapter 7), and the Worldwatch Institute. Such groups have generally relied upon the largesse of foundations and a few wealthy benefactors, rather than mass membership. The Club of Rome limited its membership to one hundred individuals. More broadly based was the Global Tomorrow Coalition of environmental and future-oriented interest groups in the United States, formed to follow up the *Global 2000 Report to the President* after 1980. Some of these groups had a large membership. But the coalition itself never played an especially visible role in mobilizing public opinion; groups could declare an affiliation to it without really integrating its efforts into their own activities.

Survivalist pressure groups have for the most part sought the ear of the powerful rather than the mobilization of any broader public (though Worldwatch publications are widely distributed). Sometimes this was successful—as for example, when President Jimmy Carter directed the

Council on Environmental Quality and the Department of State to produce *Global 2000*. Sometimes the powerful are themselves survivalists. UK Environment Minister Michael Meacher in 2003 declared that 'this is the first time in the history of the Earth that species by themselves by their own activities are at risk of generating their own demise.'[6]

The impact of discourses cannot be reduced to just the impact of interests and organizations that subscribe to them; impact may be felt in the absence of any such identifiable organization(s). Discourses take effect in largely impersonal fashion, if they can indeed manage to change the language that significant numbers of people use. So can we trace any initiatives, policies, agreements, social changes, or other phenomena to the limits discourse? Obviously coordinated population control efforts—especially in China—fall into this category. At the all-important global level, perhaps the best example may be found in coordinated international action to halt and reverse ozone depletion in the stratosphere. This is the success to which survivalists themselves point. Meadows et al. (1992: 141–60) believe that the ozone issue shows that we can indeed move at the global level 'back from beyond the limits.' Peter Haas (1992) argues that a like-minded 'epistemic community' of atmospheric scientists was the driving force behind global action, just as it should be in the survivalist discourse.

Ozone depletion was first recognized as an issue in the 1970s; the culprit was identified as chlorofluorocarbons (CFCs), chemicals found in aerosol sprays and refrigerators. Stratospheric ozone is vital in shielding life on earth from solar ultra-violet radiation, which can cause skin cancer in humans and other animals, damage photosynthesis in green plants (so threatening agriculture and forestry), and kill aquatic plankton. The issue was dramatized in the form of an 'ozone hole' appearing over Antarctica in the austral winter, identified and named in 1985. The ozone problem did, then, constitute a classic limit, which by the mid-1980s appeared to have been overshot. While global negotiations relating to this issue had been going on for some time, in 1987 there was a dramatic acceleration in the pace of global action. The 1987 Montreal Protocol was signed by twenty-four nations, covering the main CFC producers and consumers, under the auspices of the United Nations Environment Program. The Protocol committed developed nations to freeze consumption of CFCs almost immediately and impose a series of percentage cuts subsequently. Later revisions strengthened the Protocol, eventually specifying that CFCs be eliminated at the end of 1995, at least for developed countries. The rest of the world was given longer, and promised aid from developed countries to introduce substitutes and compensate for economic losses caused by lack of access to CFCs.

Meadows, Meadows and Randers (1992: 159) conclude that the ozone issue shows that 'a world government is not necessary to deal with global problems, but it is necessary to have global scientific cooperation, a global information system, and an international forum within which specific agreements can be worked out.' Litfin (1994) explains the Montreal Protocol in terms of a global discursive shift on this issue, toward a discourse of limits, or, as she refers to it, precaution. The key event to Litfin was the rhetorical force of the 'ozone hole' idea. The hole referred to seasonal and variable (though substantial) reductions in ozone concentrations over Antarctica. A 'hole' could capture the imagination in the way reams of data from monitoring stations could not. By 1987 there was actually no empirical evidence that CFCs actually damaged the ozone layer, though the required chemical reactions had been demonstrated in laboratory settings.

There is no denying that the ozone issue does represent an environmental success. However, some caution is in order before making too much of it, or seeing it as a prototype for global action on other issues, for three reasons. First, the stakes were comparatively small: CFCs are useful chemicals, but substitutes for them exist. Second, it is possible to reconstruct the history of negotiations on the ozone issue in terms not of rational solving of a collective problem (Meadows), nor of science-driven policy making (Haas), nor a discourse shift (Litfin), but rather in terms of the material interests of key actors, as Berejikian (1995) shows. The negotiations were dominated by two key actors: the United States and the European Community (EC). The United States already had legislation restricting CFCs at the national level. Corporations such as Du Pont feared that lack of any global controls on CFCs would put them at a competitive disadvantage, as they were devoting resources to the development of substitutes for CFCs. (Meadows et al. (1992: 159) praise the role played by 'flexible and responsible corporations' such as Du Pont.) So an international agreement was desirable from the point of view of Du Pont and, by extension, the United States. The EC initially dragged its feet on the issue. However, once it became clear that the United States might ban imports from the EC of products containing CFCs, the EC was much more willing to negotiate. On the ozone issue, it was fortuitous that the material interests of key players could eventually be brought into line with global environmental concerns. This coincidence should not be expected as a general rule.

Third, there is no guarantee of universal compliance with the Montreal Protocol and its subsequent strengthening. There was little problem in the developed countries delivering on their promises. Less developed countries, notably China, dragged their feet on their proposed gradual phase-out of CFCs. They found allies in the United States Congress, where an anti-environmental but, more significantly, penny-pinching Republican majority saw the finance of this phase-out by developed countries such as the United States as an undesirable monetary burden. By the late 1990s India was going full steam ahead on CFC production, and there was an emerging international black market in CFCs.

If ozone layer protection is the most prominent (qualified) success in addressing a global limit, climate change is to date the most prominent failure. The economic stakes on this question are much higher, as the foundation of industrial economies in fossil fuels is at issue. Climate change resulting from increased concentrations of greenhouse gases (notably carbon dioxide and methane) in the atmosphere promises greater frequency of extreme weather events. However, the kind of catastrophe over-dramatized in the 2004 film *The Day After Tomorrow* is less important than chronic and insidious changes: gradually rising sea levels, slow melting of the permafrost underlying tundra in Arctic regions, changes in rainfall patterns. Substantial uncertainty remains concerning the magnitude of these effects, though there is scientific consensus on the existence of climate change. The Intergovernmental Panel on Climate Change set up by the United Nations General Assembly in 1988 acts as a research sponsor and informational clearing house for scientific studies. The skeptical science denying climate change is financed by corporations with a financial interest in denial, and as such lacks credibility.

The Kyoto Protocol of 1997 on climate change ought to have been a landmark equivalent to Montreal ten years earlier. Kyoto committed developing countries to reducing their carbon dioxide emissions to 5.2 per cent below 1990 levels by 2010. Developing countries (including emerging industrial giants China and India) were given free passage. In 2001 the United States, by far the world's largest emitter, withdrew from Kyoto on the explicit grounds that US economic interests were more important than global environmental protection. But even the countries remaining inside the Kyoto agreement made little effort actually to reach its targets. Even if those targets were to be met, the impact on climate change is debatable, given the increasingly small proportion of global emissions that the Kyoto reductions represent. In contrast to the coordinated governmental action central to survivalist political philosophy, the best chance for more thoroughgoing action may come from giant insurance companies alarmed at the prospect of extreme weather events wiping out their profits (Paterson, 2001). These companies are in a position to promote climate-friendly action by their clients, as well as attaching strings to the huge investments they control.

Other possibilities for action on the climate change issue may come with a linkage to environmental security. A 2003 study commissioned by the US Department of Defense warned of famine, drought, energy shortages, and coastal flooding that could even drown cities within two decades (Schwartz and Randall, 2003). The anticipated consequences included riots, civil conflicts, nuclear brinkmanship, and political instability conducive to terrorism and war. Completely out of step with the Promethean viewpoint of the George W. Bush administration, the report was initially suppressed. Were the link to security accepted, the policy consequences would remain unclear. If security is interpreted in national terms, the result could be unilateral defensive action (resembling Garrett Hardin's 'lifeboat') rather than coordinated multilateral global action.

Survivalism: an assessment

The problem of limited compliance on the part of some nation-states on the ozone issue and the more widespread lack of national action on climate change illustrates what is perhaps the biggest challenge confronting the limits discourse. In this discourse, agency is for elites, and most importantly for elites operating on a coordinated global basis. The slogan 'Think globally, act locally' is a frequent exhortation in environmental circles; but for the discourse of limits and survival, the appropriate slogan is 'think globally, act globally.' Though the ozone issue illustrates some potential for coordinated global action (for discussion of other, less dramatic, cases, see Haas et al., 1993), as things stand the requisite global authority is missing. One survivalist—Garrett Hardin—believes it implausible, such that survivalism should be pursued within 'lifeboats' in the developed world only. Other survivalists advocate stronger global authority, but give few ideas as to how it might actually be brought into being. All kinds of political and economic currents now point in the opposite direction. The World Trade Organization is perhaps the strongest body for global governance yet established. Though sustainable development appears in the preamble to its constitution, and it has a Committee on Trade and the Environment, in practice it subordinates environmental values to economic ones. Within free-trade regimes, nation-states are compelled to pursue policies to encourage footloose investors to locate capital in their countries rather than their rivals, and this means removing environmental restraints on trade. Politically, liberal democracy is the most popular model. This model is a far cry indeed from the centralized authoritarian forms favored by the more austere survivalists, though real-world liberal democracies are not necessarily so far from the more pervasive elitism of the survivalist discourse.

Survivalism does not, then, sit easily in the real world. It also has to struggle with some powerful competing discourses. Rome is home to a Church as well as a Club. The Roman Catholic Church has consistently opposed any efforts to control human population, inasmuch as such efforts countenance contraception and abortion. The Church played an obstructive role at the 1994 United Nations International Conference on Population and Development in Cairo, in unholy alliance with Islamic fundamentalists and mostly Protestant fundamentalists from the United States. From the left, Marxists, who remember Marx's own opposition to Malthus, ridicule a discourse of limits, especially when it is propounded by powerful industrialists, and financed by corporate foundations. No less than in Marx's own day, a recognition of limits gets in the way of dreams of a future of communist abundance. Only recently has this scornful tendency among Marxists been questioned by the rise of eco-Marxism, which sees in ecological crisis the possibility of an ally that would hasten the global crisis of capitalism.

From the left more generally, opposition to survivalism comes from those who see talk of population as inherently racist (for the United States, see Chock, 1995; for Australia, see White, 1994). An ontology that stresses 'population' and the denial of agency to members of that population looks, from this left/multicultural direction, as if it were designed to control and discriminate against those ethnic groups whose numbers are increasing most rapidly. In the United States, these are generally nonwhites and especially Hispanics, whose numbers are bolstered by immigration both legal and illegal. In 2004 a group opposed to immigration attempted to gain control of the Sierra Club's governing board, but was defeated in a membership election.

An assault on the discourse of limits has also come from ecofeminism. Ecofeminists affirming pre-patriarchal symbols of fertility contend that population control means control of women by a male power structure (Diamond, 1994), in the interests of producing the kinds of consumers required by global capitalism (Sandilands, 1999: 88). For women to regain their places in harmony with a living, fertile earth it means breaking free from all patriarchal shackles—including population control experts and their political masters. The survivalists do give them plenty of ammunition here, as for example when Garrett Hardin (1993: 258) states that 'we need to devise acceptable ways of influencing the desires of women in the light of community needs.'

The religious, Marxist, left multiculturalist, and ecofeminist arguments against the discourse miss their target. Religious arguments reduce to dogma concerning fundamental tenets of faith for individual (micro) behavior combined with insensitivity to how these tenets play out at the macro level. Orthodox Marxists, now thin on the ground, disapprove of the limits discourse because of its inconvenient implications, not because they can fault its logic. Left multiculturalists, in exposing racist and sexist aspects of limits discourse, do not thereby solve the problem of population pressure; they sweep it under the carpet, or implicitly assume that the interests of particular ethnic groups must always trump any global interest. In other words, they advocate maintenance and expansion of a commons in the interests of social justice. They actually have nothing to say about the reality or otherwise of limits, beyond dismissing them as social constructions that serve established political power – which is obviously false, given that global capitalism is so much better served by Promethean discourse that denies limits.

A better deployment of social justice concerns here would begin by noting that it is the prosperous peoples of the world who impose the greatest burden on the ecosphere. Further, starvation in the Third World results not from global food shortages, but structures and processes that favor the profits of the few over the basic needs of the many (Lappé and Collins, 1977).

The counterclaims of these competing discourses cannot easily stand scrutiny—indeed, they do not take on the limits discourse in any terms that could engage the problems it has identified. But these counterclaims help explain the real-world fate of survivalism and limits.

There does exist a discourse that has engaged survivalism more directly, and on ground where arguments can be made, as opposed to dogma asserted. This opponent is Promethean, and its defining feature is the denial of limits. Given that it is rooted in industrialism, but only makes sense as a reaction against the limits discourse, it is appropriate to discuss this Promethean discourse and the challenge it presents to survivalism at length in the next chapter.

The limits discourse has, then, had limited impact in its own terms. Recall that survivalism seeks coordinated, central action, with foresight built in. Very little such action is apparent. The time horizons of governments and corporations remain short. The emerging global political economy is thoroughly inhospitable to survivalist concerns. The limits discourse is in evidence at the periodic United Nations conferences on population and the environment (notably the Conference on the Human Environment in Stockholm in 1972, the Conference on Environment and Development in Rio in 1992, the International Conference on Population and Development in Cairo in 1994, and the World Summit on Sustainable Development in 2002), though increasingly these gatherings are dominated by the competing discourse of sustainable development. Eventually we find survivalists lamenting time wasted as a result of prior warnings unheeded—which is why Meadows et al. (1992) believe the world has now gone 'beyond the limits.' Garrett Hardin (1993) bemoans the fact that the population issue was not on anyone's agenda for Earth Day in 1990, mostly, he believes, as a result of the fear of the corporate sponsors of Earth Day of offending the anti-limits groups I have mentioned (1993: 3).

One other factor that can restrict the impact of survivalism is a silence about environmental problems as they arise at local, regional, and national levels. Survivalism is about both thinking and acting globally. Some of the other discourses have more to say about environmental issues at these other levels. However, it is conceivable that one could be a survivalist for global issues, while subscribing to some other discourse when it comes to local issues.

Nothing in this patchy record means that the survivalist discourse is erroneous in its stress on limits, and it may yet find vindication if, for example, the global greenhouse effect comes to the worst. Politically, there has never been enough imagination as to how the agenda might be pursued, and the discourse never really got past the simplistic draconian authoritarianism of the 1970s survivalists. Survivalists do not quite know what to say or do about global capitalism, especially given that some of their financial sponsors are global capitalists. These problems notwithstanding, the impact of the discourse of limits and survival should not be dismissed. If nothing else, it raised the stakes in the establishment of the environment as a key issue, perhaps the key issue for the twenty-first century.

NOTES

1 His analysis was also formally identical to the account of the seventeenth-century political philosopher Thomas Hobbes of the 'war of each against all.'

2 The Club's founder and president, Aurelio Peccei, eventually published his own survivalist tract, *One Hundred Pages for the Future* (1981). This book is actually 187 pages long, at least in the edition I have, surely a case of growth beyond the limit.

3 Countries such as the United States, in contrast, require relatively restrained policies, such as heavy taxes on diapers and toys to discourage people from having children. Later, Paul and Anne Ehrlich (1974) give up on the possibility of any government doing much right, and propose only that individuals prepare themselves for the coming crunch by laying in stores of food and practicing self-sufficiency. With this suggestion, the Ehrlichs find common ground with the rightwing, gun-toting survivalists in the United States. But such aggressive individualism is generally absent from the environmental discourse of limits and survival.

4 Later, Ophuls protests that he raises the specter of authoritarianism simply as a warning of what might have to happen unless humanity gets its political-ecological house in order through less draconian means (Ophuls and Boyan, 1992: 312). In addition to this authoritarian model, his book contains, in uneasy juxtaposition, a decentralized, Jeffersonian political economy of self-reliant small-scale communities.

5 *Guardian*, London, February 14, 2003.

6 *Guardian*, London, February 14, 2003.

3

Growth Forever: The Promethean Response

The Promethean background

Discourses do not need conscious articulation. They can be so ingrained and taken-for-granted that it would never occur to anyone to mention them. (Analogously, most speakers of the English language could not articulate the basic principles of grammar and syntax they use every day.) Such was long the case for the environmental discourse which can, now that it has been articulated, be styled Promethean. In Greek mythology Prometheus stole fire from Zeus, and so vastly increased the human capacity to manipulate the world. Prometheans have unlimited confidence in the ability of humans and their technologies to overcome any problems—including environmental problems.

The term 'cornucopian' is sometimes associated with this denial of environmental limits. Cornucopia means abundant natural supply: unlimited natural resources, unlimited ability of natural systems to absorb pollutants, and unlimited corrective capacity in natural systems. However, Julian Simon protests that this is a misnomer: 'The school of thought that I represent here is not cornucopian. I do not believe that *nature* is limitlessly bountiful. . . . our cornucopia is the human mind and heart not a Santa Claus natural environment' (1981: 41). Simon's protests notwithstanding, I will show that members of this school, including Simon himself, do portray a Santa Claus natural environment at key junctures. So strictly speaking this discourse should be styled Promethean/cornucopian. But that is too much of a mouthful, so let me just call it Promethean, which really does capture the essence better than 'cornucopian.'

For several centuries, at least in the West, the dominant Promethean order had been taken for granted. The Industrial Revolution produced technological changes that made materials close to home (such as coal and later oil) into useful resources. At the same time, European colonial expansion opened up new continents and oceans for exploitation. Capitalist economic growth became taken as the normal condition of a healthy society. Even those who looked forward to a future beyond capitalism, notably Karl Marx, applauded technological progress, economic growth, and the conquest of nature.

The power of unarticulated Promethean discourse is still felt today, as just about every government sees its first task as promoting economic growth. The entire way in which economic news is reported assumes that growth is good. This refers to growth in wealth, growth in income, growth in profits, growth in the stock market, growth in employment, growth in housing starts, growth in passenger miles traveled. That economic growth usually means increased stress on environmental systems—more pollution, more congestion, faster depletion of resources—is never reported along with these economic aggregates (though this stress is reported elsewhere). The political-economic discourse of liberal capitalist systems still generally floats free from any sense of environmental constraints. 'Economy' and 'environment' are put into different boxes.

Promethean argument to the foreground

The rise of survivalism described in the previous chapter meant that Promethean discourse had to be articulated and defended, rather than just taken for granted. Economists are at the forefront of the Promethean counterattack. The economists' basic argument was established as early as 1963 with the publication of *Scarcity and Growth* by Harold Barnett and Chandler Morse, produced under the sponsorship of the Washington think-tank Resources for the Future. Economists have always said that price is a measure of scarcity: if the real price of a good goes up, that means demand in relation to supply is increasing. Conversely, if the price of a good falls, then demand relative to supply is falling. This logic can be applied to the goods we call natural resources. Barnett and Morse gathered long-term trend data for the prices of a number of 'extractive goods': agricultural products, minerals, fisheries products, and timber. In every case except forest products the story was the same. Barnett and Morse showed that since at least the beginning of the twentieth century, the real price (i.e. after adjusting for inflation) of natural resources had been falling. If price measures scarcity, this means natural resources are becoming more abundant with time. Updates of the Barnett and Morse analysis continued to tell a similar story (Smith, 1979; Taylor, 1993).

When the limits to growth argument and associated survivalist discourse arrived with such a bang in the early 1970s, Promethean economists did, then, have a ready-made argument and plenty of data to throw back at the survivalists. Wilfred Beckerman (1974) deployed long-run evidence from price trends against the *Limits to Growth* global modelers to argue that there was nothing wrong with projecting economic growth into an indefinite future. It became the standard criticism of the *Limits* modelers that their computer models did not include any role for either prices or technology. Twenty years later, Beckerman saw no difficulty in deploying the same kind of evidence against another generation of survivalists to argue that *Small is Stupid* (Beckerman, 1995; the title's allusion is to E. F. Schumacher's *Small is Beautiful*). Later still, his *A Poverty of Reason* (Beckerman, 2000) took aim at sustainable development as well as limits. Julian Simon put his money where his mouth was in a famous bet with survivalist Paul Ehrlich in 1980. Simon bet that the real price of any set of natural resources that Ehrlich cared to name would be lower at any time in the future than in 1980. Ehrlich specified copper, chrome, nickel, tin, and tungsten, with 1990 as the date. Come 1990, the price of copper was 24 per cent lower than in 1980, chrome 40 per cent lower, nickel 8 per cent lower, tin 68 per cent lower, and tungsten 78 per cent lower. Ehrlich sent Simon a check for $1,000 (see Michaels, 1993: 368–9). Later, Simon would renew his challenge in the context of a debate with survivalist Norman Myers; he was so confident that he increased the offer to a month's pay, and extended it to any measure of human welfare in any country or region of the world (Myers and Simon, 1994: 20–1, 115; see also Simon 1996: 33–6). This time, Simon found no takers.

Why do the prices of natural resources keep falling, suggesting increasing abundance? If a shortage threatens, there is money to be made in either finding new sources of the resource in question, or in developing substitutes. In this light, there is nothing new about resource scarcity, or about response to it. In sixteenth- and seventeenth-century Europe, wood was the key energy resource, and so when wood supplies seemed to be running out an energy crisis looked imminent. The response was the development of coal as a fuel, which in turn made the technologies of the Industrial Revolution possible—and coal itself came to be mined more cheaply using these technologies (Nef, 1977). Come the mid-nineteenth century, the economist William Jevons (1865) predicted that coal would soon run out, and that the wheels of British industry would stop turning. He need not have worried; not only were more deposits of coal continually discovered, but oil was soon developed as an energy resource.

Economists such as Barnett, Morse, Beckerman, and Simon see no problem in projecting such happy trends in resource and energy availability and price into the future. Just how far into the future? 'We expect this benign trend to continue at least until our sun ceases to shine in perhaps 7 billion years, and until exhaustion of the elemental inputs for fission (and perhaps for fusion)' (Simon and Kahn, 1984: 25). No modesty here! Those not convinced that economists can forecast inflation and unemployment over the next year or so might be a bit hesitant about accepting the seven-billion-year forecast.

In the 1980s Julian Simon established himself as the leading American Promethean, and broadened the argument beyond resource prices to encompass trends over time of indicators of human wellbeing such as life expectancy, food supply per capita, amount of arable land, air and water quality, amount of parkland, forest cover, and fish catch. The indicators he sought were mostly global, though national and regional data were adduced too. Life expectancy plays a key role for Simon as a surrogate for pollution. He allows that with time some forms of pollution increase as others decrease; so the introduction of the internal combustion engine saw an increase in pollutants such as carbon monoxide and ozone, but a massive decrease in the horse droppings in which city streets were often quite literally awash. What matters, Simon says, is the net overall effect of offsetting pollution increases and decreases on human health—and the best summary measure of that is life expectancy (1981: 130–1). The long-term trend evidence is that in all parts of the world people are living longer; therefore in all parts of the world pollution is falling. In 1995 Beckerman observed that over the previous thirty years life expectancy globally had risen from 53 to 66 years (1995: 111). (Later, however, AIDS would cause falling life expectancy in much of sub-Saharan Africa; and in the 1990s life expectancy declined in Russia.) Easterbrook (1995) declares that 'In the Western world, the Age of Pollution is nearly over.' Bradley (2003) argues that all the pollution issues associated with energy production have now been solved (though climate change resulting from fossil fuel use may still require a bit of attention).

Simon is not always as scrupulous as he might be in the kinds of evidence he adduces. For example, in 1984 he tried to show that the United States was becoming less crowded through reference to a graph showing a massive increase in the amount of land in national parks over the 1950–80 period (Simon and Kahn, 1984: 8). In fact, almost the entire increase occurs in 1979. What happens in 1979, which Simon fails to mention, is passage of the Alaska National Interest Lands Conservation Act, which for the first time classified federal government lands in Alaska. Some of these lands were classified as national parks. Many are accessible only by bush plane. One can imagine the sighs of relief in 1979 echoing around the Bronx, South Central Los Angeles, and Chicago, as suddenly their residents all felt less crowded.

Julian Simon died in 1998. His place as the public face of Prometheanism was soon taken by Bjørn Lomborg. Lomborg is a political scientist and statistician rather than economist, and so says little about the underlying economic mechanisms that alleviate scarcity. He focuses narrowly on the trends themselves, and reaches exactly the same conclusions as Simon on global aggregates (though like Simon he is selective in his level of analysis, using regional data when it suits his purposes). Lomborg made a splash in 2001 with the publication of *The Skeptical Environmentalist* ('Environmentalist' because he claims to be a former Greenpeace member). Extracts and summaries of the book quickly appeared in the *New York Times, Guardian* (London), and *The Economist.* The claims echoed Simon: natural resources, energy, and food are becoming more abundant, fewer people are starving, life expectancy is increasing, pollution is eventually reduced by economic growth, species extinction presents a limited and manageable problem, forests are not shrinking.

Lomborg met a storm of criticism from environmentalists and scientists (in some cases those whose work he drew on in the book).[1] Environmentalists (e.g., Burke, 2001) objected that Lomborg begins with a caricature of environmentalism. In Part I of his book, Lomborg sets out an alleged 'Litany' of environmentalism:

> The environment is in poor shape here on Earth. Our resources are running out. The population is ever growing, leaving less and less to eat. The air and water are becoming ever more polluted. The planet's species are becoming extinct in vast numbers . . . the forests are disappearing, fish stocks are collapsing. . . . We are defiling our Earth, the fertile topsoil is disappearing, we are paving over nature, destroying the wilderness, decimating the biosphere and will end up killing ourselves in the process. The world's ecosystem is breaking down. We are fast approaching the absolute limits of viability, and the limits to growth are becoming apparent. (Lomborg, 2001*a*: 4).

Lomborg goes on to say about the Litany: 'There is just one problem: it does not seem to be backed up by the available evidence.' But there is in fact another problem: hardly any environmentalists actually subscribe to it. The Litany captures only an extreme position in the survivalist discourse, more popular in the early 1970s than in the early 2000s. Lomborg's main targets are the Worldwatch Institute and the early works of Paul Ehrlich (which pre-date *Global 2000*, Simon's main target). Lomborg fails to recognize the variety of environmentalist positions—not least one that would credit environmental improvements of the kind he charts to the political efforts of environmentalists influencing public policy.

Natural scientists objected that Lomborg presented selective and distorted interpretations of their data. A special issue of *Scientific American* (January 2002) was devoted to debunking Lomborg. Aside from *ad hominem* remarks about a political scientist trespassing on scientific territory, the authors criticized Lomborg's misinterpretation of scientific sources, selective presentation of evidence, stress on the quantity as opposed to quality of resources such as forests (equating old growth ecosystems with timber plantations), and insensitivity to uncertainty about complex systems. Charges against Lomborg were taken to the Danish Committee on Scientific Dishonesty, which found against him in that 'the objective criteria for scientific dishonesty have been met,' but excused him on the grounds he didn't know what he was doing.[2] In December 2003 the Danish Ministry of Technology overturned the Committee's negative verdict, by which time Lomborg had been appointed Director of Denmark's Environmental Assessment Institute by a right-wing government under a prime minister to whom Lomborg had access. This episode reveals much more about the peculiar politics of science in Denmark than it does about the veracity or falsity of Lomborg's analysis.

Analysis of Promethean discourse

Basic entities whose existence is recognized or constructed

For Prometheans, natural resources, ecosystems, and indeed nature itself, do not exist. This denial can explain just about everything there is worth knowing about Promethean discourse. This claim about nature's non-existence, at least as anything more than a store of matter and energy, might seem startling. Consider, though, Simon's basic argument about natural resources. He affirms, time and again, that the supply of natural resources is infinite. Why? Because there is no fixed supply of resources: 'Resources are only sought and found as they are needed' (1981: 40). Thus there is no point in measuring the quantity of reserves remaining: if more is needed, more will be sought and found. But just what are these things we call natural resources anyway, if, as one Promethean argues, 'Not a single natural resource has ever been created by "nature" ' (Taylor, 1993: 378)? The answer is that 'natural' resources are created by humans transforming matter. Nature is, indeed, just brute matter; and in their wilder moments (as in the Simon and Kahn seven-billion-year projection mentioned earlier) Prometheans believe matter is infinitely transformable, given enough energy. The medieval alchemists believed base metals could be turned into gold. Prometheans believe that with enough energy iron can be turned into copper (Myers and Simon, 1994: 100n)—which indeed it can be, though the amount of energy required is massive. Similarly, deserts can be turned into cropland, outer space can be colonized.

With enough energy, with the fruits of economic growth, we can also take care of pollution (see Lewis, 1992: 184). As Beckerman (1995: 25–6) puts it, 'if you want a better environment in general and, in particular, reasonable access to clean drinking water, adequate sanitation and an acceptable urban air quality, you have to become rich.' Lomborg echoes this sentiment: 'only when we are sufficiently rich can we afford the luxury of caring about the environment' (2001*a*: 33). Pollution is, in Promethean light, just matter in the wrong place in the wrong form, and with enough skilled application of energy, that can be corrected.

Nature does not, then, exist as anything more than brute matter. Though Prometheans might occasionally use the word 'ecosystem,' the concept plays no part in their discourse, in which ecosystems do not constrain human activity. Accordingly, for Prometheans 'the term "carrying capacity" has by now no useful meaning' (Simon and Kahn, 1984: 45). Lomborg's position on limits is a bit different from Simon's: he accepts that when it comes to fossil fuels and minerals, 'there must be some limits to the amount . . . that can be extracted', but 'That limit is far greater than many environmentalists would have people believe' (2001*b*).

Having dealt with absences from the Promethean ontology, what basic entities does it recognize or construct? In short: people, markets, prices, energy, technology. Prometheans talk a lot about population—Julian Simon became famous mostly as a result of his intervention in debates over global population, in which he celebrated population growth. But population is not constructed in the way it is for survivalists, as an aggregate entity to be controlled. I will return to the implications of this difference shortly.

Assumptions about natural relationships

I have already noted that the Promethean discourse comes close to denying the very existence of nature, which is at most seen in inert, passive terms. The most important natural relationship taken for granted is therefore a hierarchy in which humans (and in particular human minds) dominate everything else. This domination does not need to be organized, or consciously maintained; it just exists. In their more extreme moments, Prometheans believe that a total control of nature is within our grasp (for an early explicit statement, see Murphy, 1967).

Beyond human domination, the other relationship seen as natural is competition between humans, through which innovative means for overcoming emerging scarcities can best be generated. So when the Organization for Petroleum Exporting Countries organized oil embargoes in the 1970s it was competition which spurred the search for non-OPEC sources of oil, and for cars that would use less gasoline. This emphasis reveals an affinity between the Promethean discourse and proponents of the market. Prometheans see little need for government to do much in the way of environmental and natural resources policy: if long-term trends are improving, the best thing government can do is leave well alone. Inasmuch as they attend to government, Prometheans see mostly sources of ill. As the Promethean physicist Bernard Cohen (1984: 566) puts it in reference to the United States, 'Given a rational and supportive public policy, science and technology can provide not only for the twenty-first century but for ever.' But, he avers, the necessary support for nuclear energy and other constructive endeavors is missing. For:

> our government's science and technology policy is now guided by uninformed and emotion-driven public opinion rather than by sound scientific advice. Unfortunately, this public opinion is controlled by the media, a group of scientific illiterates drunk with power, heavily influenced by irrelevant political ideologies, and so misguided as to believe that they are more capable than the scientific community of making scientific decisions. (p. 566)

Agents and their motives

In the Promethean discourse, agency—the capacity to act—is for everyone, mainly as economic actors. People going about their business, pursuing their selfish interests, will together ensure a bright environmental future. This is an application to resource and environmental issues of the 'invisible hand' working in the market system, first celebrated by Adam Smith in the late eighteenth century.

As seen in the preceding chapter, survivalism denies agency to populations, which are treated as problems to be controlled. Promethean discourse, in contrast, celebrates the people who compose populations. If individuals are problem solvers, all potentially contributing to the betterment of humanity's lot, then the more people the better. In denying the existence of limits, Prometheans generally also deny the need to worry about rising populations, be it on a national, regional, or global scale. Julian Simon, in particular, pointed out that rising populations have been accompanied by rising, not falling, life expectancy, and increased, not reduced, income per head. Certainly this is true at the global level: the same decades which have seen population explosion have also seen life expectancy rising to historically unprecedented levels on a global scale, and the prices of natural resources continuing to fall. Individuals, Julian Simon believes, make good decisions about the number of children they have, such that 'population size adjusts to productive conditions rather than being an uncontrolled monster' (1981: 162–3). For Simon, 'the ultimate resource is people—skilled, spirited, and hopeful people who will exert their wills and imagination for their own benefit, and inevitably they will benefit not only themselves, but the rest of us as well' (1996: 589). If people are good, then more people will always be better. Implicit here is the idea that the supply of ingenuity is proportional to the number of people (an assumption questioned by Homer-Dixon, 2000, who suggests limits to this supply).

Not all Prometheans share Simon's cavalier position on population. Easterbrook (1995) calls for world population stabilization in the short term, while allowing that in the long term the Earth could support several times its current population. Beckerman (1995: 63, 173–4) allows that developing countries have a population problem. He believes (as does Lomborg, 2001*b*) that the world's most pressing environmental problems are associated with the poor in these countries lacking access to clean water and good sanitation. The obvious solution is for them to become rich, and that may be easier the fewer of them there are.

Key metaphors and other rhetorical devices

The key Promethean metaphor is mechanistic. Machines are constructed from simple components—ultimately, simple resources—through the application of human skill and energy. Thereafter they do useful things. A solution to any kind of problem can be pieced together in like manner for Prometheans, be it the restoration of malfunctions in the human environment (such as pollution or wilderness destruction), or the creation of natural resources for human use.

The main weapon in the Promethean rhetorical arsenal is the trend. Prometheans are at their happiest when presenting graphs depicting declining resource prices, increasing parklands, croplands, and forests, increasing life expectancy, increasing crop yields and fisheries catches, and so forth. The explicit accompanying message is: the trend can be extrapolated indefinitely into the future—more than seven billion years into the future, as we have seen for Simon. Lomborg is slightly more cautious, saying that 'trends provide the best information about how things have progressed and are likely to progress' (2001*a*: 6). Note how different this is from the survivalist modeling of the interaction of different variables (such as population growth, resource use, and environmental damage). The trends presented in graphs or figures by Prometheans are single-variable, and no attempt is made to model interactions. Prometheans would say that the interactive models of survivalists are inaccurate, simplified, and speculative. The difference underscores the Promethean neglect of the existence of eco*systems*, in which by definition many factors interact.

The impact of Promethean discourse

Promethean discourse flourished alongside capitalism and the Industrial Revolution, with its unbounded faith in the ability of humans to manipulate the world in ever more effective fashion. Such was human progress. Thus the first place to look for the impact of the discourse would be in our dominant institutions: a capitalist economy geared to perpetual economic growth, and a political system whose main task is to facilitate the conditions for that growth. Discourse and institutions co-evolved. When it comes to political institutions, the Promethean discourse constitutes much of their software, if the hardware is composed of formal laws and constitutions. That is, institutions of government such as parliaments, executives, and bureaucracies require sets of understandings shared by the people who work within them in order to coordinate their operations. The main shared understanding in the capitalist democracies has long been that growth is good.

BOX 3.1 **Promethean discourse analysis**

1. **Basic entities recognized or constructed**
   - Nature as only brute matter
   - Markets
   - Prices
   - Energy
   - Technology
   - People
2. **Assumptions about natural relationships**
   - Hierarchy of humans over everything else
   - Competition
3. **Agents and their motives**
   - Everyone; motivated by material self-interest
4. **Key metaphors and other rhetorical devices**
   - Mechanistic
   - Trends

Once the environmental challenge arrived in the late 1960s, and especially with the terms of debate set by survivalism in the early 1970s, the Promethean discourse was very much on the defensive, and so pressured to articulate its key tenets for the first time. These newly articulated tenets eventually found a ready and sympathetic ear in the form of President Reagan and his associates, and later in the presidency of George W. Bush. These presidencies are the high points of influence of Promethean discourse on policies and institutions.

Reagan's anti-government, market-oriented ideology was quite clear in its enthusiasm for economic growth, and that the main impediment to growth was excessive governmental regulation. Environmental regulation was a particular target. Candidate Reagan, in his acceptance speech to the 1980 Republican Party convention, declared that 'the economic prosperity of our people is a fundamental part of our environment'. A moment's thought will reveal that this is a meaningless sentence; but its rhetorical importance is to declare that environmental goals should be subordinated to economic ones. Just how this would happen became evident as soon as Reagan entered the White House in 1981 (for details, see Vig and Kraft, 1984).

Two key appointments symbolized the Reaganite Promethean approach. James Watt was appointed Secretary of the Interior, responsible for overseeing the vast bulk of federal lands. Anne Gorsuch (later Burford) was appointed Administrator of the Environmental Protection Agency (EPA), charged with administering the nation's anti-pollution policy. Both Watt and Gorsuch Burford were hostile to most of the legislation they were supposed to administer. Watt was by background a 'Sagebrush Rebel,' identified with a movement in the rural West of the United States keen to transfer ownership of federal lands (including national parks and wilderness areas) to the states, hoping that the states would open these areas to loggers, miners, and ranchers. Though their rhetoric was anti-government, Sagebrush Rebels were not especially interested in the free market; they did not want to pay market prices for access to land, timber, or minerals, but rather sought a continuation of heavy government subsidy of these activities, with privilege accorded to established users (such as holders of grazing leases on public land) rather than open to the highest bidder. In this, the Sagebrush Rebels differed from those Prometheans who stress markets and private property. Still, Watt's view was clearly Promethean:

resources were there to be used for human benefit, not locked away. He characterized environmentalists as Nazis or Bolsheviks, and anyone who did not share his views as un-American. In 1990 he suggested that 'if the troubles with environmentalists can't be solved in the jury box or the ballot box, perhaps the cartridge box should be used' (quoted in Dowie, 1995: 97). Watt's three years as Secretary of the Interior (he was forced to resign over a racist and offensive joke about the composition of a review committee) were turbulent, as he was in constant battle with a congressional majority. In addition, he proved the best recruiting agent that the environmental movement ever had. Thus the massive policy changes he sought never materialized: nobody could galvanize the environmental opposition into action quite like James Watt.

Except perhaps for Anne Gorsuch Burford. Burford attempted to purge the EPA of individuals who actually believed in the agency's mission, and turned policy making over to the polluters the EPA was supposed to regulate. The involvement of the chemical industry in running EPA's hazardous waste program was eventually shown to have crossed the bounds of legality. Burford, too, clashed repeatedly with Congress, and was forced to resign under fire in early 1983. She left behind an agency with its budget and personnel slashed, its morale shattered, and its mission compromised. Her assistant administrator for hazardous waste, Rita Lavelle, received a six-month prison sentence.

Of these two environmental nemeses, Burford had the greater impact. Watt had minimal effect, mainly because the checks and balances built into the United States system of government stymied his attempts at radical reform. He simply could not manage the federal lands the way he wanted to. Burford, in contrast, was able to paralyze some essential EPA regulatory functions, and damage its enforcement capabilities. However, the damage proved not to be lasting: Burford was replaced as Administrator by William Ruckelshaus, who had been the first Administrator of the EPA under President Nixon. Ruckelshaus began the process of restoring the EPA to its traditional mission. Yet the Reagan administration continued to operate under an Executive Order (number 11291) issued in February 1981 which specified that economic criteria should take priority in the formulation of rules and regulations by government agencies.

The later Reagan years saw a retreat from some of the excesses (and flamboyance) of the Watt–Burford era. No longer was the Promethean agenda in the hands of clowns and criminals at the highest level of government (criminality being evident at the EPA). Yet Promethean discourse set the tone for the rest of the 1980s in United States policy making. Nowhere was this more evident than in US actions in the international arena, where wholesale reversal of US commitment to international environmental governance took place (see Caldwell, 1984). The United States withdrew support and finance for international treaties or programs concerning the law of the sea, transboundary air pollution (especially acid rain), trade of nuclear materials, and the United Nations Environment Program. Given the weight of the United States in international affairs, such a stance effectively blocked coordinated international environmental policy. Of course, Prometheans would argue that no such policy is necessary. Consistent with this view, the United States also ended support for international population control programs. Influential here was not just the Promethean argument favoring population growth, but also the objections of the Christian Right and the Catholic Church to programs that involved abortion as an option. In the 1980s, the only real exception to US foot-dragging on international environmental affairs came with the issue of ozone layer depletion, discussed in the previous chapter.

The influence of Promethean discourse in US policy in the 1980s can be discerned quite clearly in the person of Julian Simon. Simon himself rejected the 'spaceship earth' metaphor that appears in survivalist discourse; otherwise, one could describe him as spaceship earth's first science officer in the 1980s.[3] Simon and like-minded futurist Herman Kahn were originally scheduled to conduct the federal government's interagency review of the survivalist *Global 2000* report produced in the waning months of the Carter presidency. That proposal was blocked by Alan Hill, Chair of the President's Council on Environmental Quality, and Simon and Kahn instead produced their response with financial support from the right-wing Heritage Foundation. This response was published in 1984 as *The Resourceful Earth* (Simon and Kahn, 1984), one of the most visible and influential Promethean documents. The Heritage Foundation played a key role in formulating policy initiatives for the Reagan administration.

Come the late 1980s, the power of Promethean discourse in US environmental affairs receded. George Bush the elder distanced himself from his predecessor on environmental policy, declaring in 1988 that he wanted to be 'the environmental president.' Only on the population issue did extreme Prometheanism hold its power—largely because of the common cause Prometheans could make with the Christian Right and the Catholic Church. With Republican gains in Congress in 1994 led by House Speaker Newt Gingrich, this natalist alliance gained new ground; and proposals dormant since the Watt–Burford years of Promethean excess were once again heard in Washington. In the early 1980s it had been Congress which had blocked these excesses; come 1995, it was Congress, or at least the Republican majority, pushing for weaker anti-pollution laws, the end of the Endangered Species Act, and opening federal lands to loggers, miners, and ranchers. At the same time, the US West saw an echo of the Sagebrush Rebellion in the form of the Wise Use Movement, whose goal again was to remove control of federal lands to the state and local level. The Wise Use Movement combined populist rhetoric with funding from resource industry corporations, though it had some grassroots support in resource-dependent areas in the West (Thiele, 1999: 203–9).

Promethean discourse returned to the White House with a vengeance with the installation of George W. Bush in 2000. Though radical, the attack on environmental policy was often low-key and subtle, with no flamboyant figures to act as lightning-rods and recruiting agents for environmentalists. The EPA was initially headed by moderate Christine Whitman, Interior by the low-profile Gale Norton. The best hate figure for environmentalists was Vice-President Dick Cheney, who brought the values of the unreconstructed wing of the oil industry into government. Cheney headed a task force on energy policy composed of executives from the oil, coal, and vehicle manufacturing industries. In introducing its report in 2002, Cheney declared that 'conservation may be a sign of personal virtue, but it is not a sufficient basis for a sound, comprehensive energy policy.' 'Sound and comprehensive' meant increasing the supply of traditional energy sources, especially fossil fuels. This in turn meant promoting oil and gas development on federal lands—most controversially, the Arctic National Wildlife Refuge in Alaska—and removing restrictions on the use of fuels. The emphasis on supply over conservation could be linked to the war in Iraq, though in the short term at least the effect on oil supply was the opposite of that intended.

Promethean discourse backed US withdrawal from the Kyoto Protocol on climate change in 2001, on the grounds that US economic interests took precedence over environmental considerations—which could take care of themselves. As Wapner (2003: 7) put it, 'the hegemon has essentially checked out of the business of global environmental protection.' The administration had little to fear from a Republican-dominated Congress and weakened judicial system. Unlike the Reagan era, checks and balances failed. However, the administration still feared the environmental leanings of public opinion. So domestically, policy actions were cloaked in the language of environmentalism. Opening federal forests to (uneconomic) logging was justified by fire protection, and described as the 'Healthy Forests Initiative' (signed into law in 2003). Similarly, a weakening of air pollution controls was styled the 'Clear Skies Initiative.' In 2004 Interior Secretary Gale Norton spoke of her department's 'new environmentalism.'

Most policy changes took effect 'below the radar,' in administrative interpretation of legislation and regulation, reduced enforcement of regulation, funding cuts at EPA, and the politicization of scientific advice. Scientists were expected to toe the line and suppress unwelcome evidence and advice. Few species were listed as endangered, and few designations of critical habitat made. Some administrative decisions were profound in their effects. Examples include a 2003 decision to allow coal-burning electrical utilities to upgrade old and dirty plants without meeting current anti-pollution standards, the attempted removal of Clean Water Act protection for vast areas of 'isolated' waters, and the attempted exemption of the US military from all environmental laws (for a catalogue, see Environment 2004, 2003).

Bjørn Lomborg looked as though he could play the role for the George W. Bush administration that Julian Simon played for Reagan. In February 2004 an email sent to all Republican members of Congress recommended they reply to Democratic criticisms of Bush's environmental policy with rosy quotes from Lomborg.[4] But Lomborg proved reluctant to embrace Bush. Right-wing think-tanks continued to push the Promethean agenda in the United States (for example, Bailey's *Earth Report 2000* is linked to the Reason Foundation).

It is, then, mainly in the United States that explicit Promethean discourse has gained significant influence (though there are Promethean publicists in other countries, such as Wilfred Beckerman in the UK and Bjorn Lømborg in Denmark). This American exceptionalism was highlighted in the 1980s and after 2001. In the 1980s, the United States sometimes found itself casting the sole vote in the United Nations General

Assembly against particular environmental measures—such as the World Charter for Nature in 1982, and a motion against trade in hazardous substances in 1983. (Of course, not all of the countries voting for these measures could be said to be paragons of environmental virtue.) In 2001, the United States stood almost alone in renouncing The Kyoto Protocol (accompanied only by Australia, whose foreign policy at the time was to follow US instructions). In the United States, Promethean discourse resonates with the interests of both capitalist market zealots and Christian conservatives, not to mention miners, loggers, and ranchers accustomed to subsidized access to resources. Such constituencies are smaller or absent in other countries. But even in the United States, large corporations that might be expected to benefit from broad dissemination of a Promethean view often prefer discourses with at least a veneer of environmental concern.

Promethean discourse: an assessment

Assessment of the Promethean discourse may begin by noting that without a cornucopian adjunct, the discourse is radically incomplete, the protestations of its adherents such as Julian Simon notwithstanding. Why is this?

Prometheans believe that humans left to their own devices will automatically generate solutions to problems—and that an invisible hand guarantees good collective consequences. To substantiate this claim, they return time and again to examples such as the introduction of motor vehicles leading to cleaner and healthier city streets with the removal of horses and their droppings. But there is no guarantee that such benign side-effects of individual actions will always occur. It is not only survivalists who stress the centrality in environmental affairs of the tragedy of the commons, the essence of which is that materially rational individual decisions can produce disastrous macro-level consequences. What can Prometheans say about, for example, global warming, where millions of rational individuals contribute to the buildup of carbon dioxide in the atmosphere by burning fossil fuels? The Promethean answer is simply to deny that global warming is a serious problem (see, for example, Beckerman, 1995: 79–87; Michaels, 1993; Ridley, 1995: 21–4). Bradley (2003: 20–1) even argues that climate change is mostly benign, bringing 'warmth, moisture and carbon fertlization.' Lomborg allows that climate change is happening, but claims it is much cheaper to accept it and adjust, rather than fight it. Similarly, Bradley (2003) suggests that the best approach is to generate wealth through economic growth to finance adaptation.

The Promethean Aaron Wildavsky goes through a whole series of environmental risks to show that really the scientific evidence shows there is no cause for alarm—and happily generalizes this conclusion to all risks, including ones he has not studied (1995: 447). Similarly, Simon suggests that we extend his conclusions from cases where good evidence is available to cases where it is unavailable, such as 'the ozone layer, the greenhouse effect, acid rain, and their kin' (Myers and Simon, 1994: xvi). When it comes to loss of biodiversity through species extinction, Prometheans again deny the problem, arguing that there are no reliable statistics to show wholesale extinctions caused by human activities such as deforestation (see Simon's contribution in Myers and Simon, 1994: 40; Lomborg, 2001*a*: 255).

What is coming to the rescue of the Prometheans here? Julian Simon claims that the difference between himself and the survivalists is that he bases his arguments on evidence, whereas survivalists rely on theory (Myers and Simon, 1994: 148). But the Promethean argument can only stand with not just a theory of human ingenuity, but also a theory of nature's abundance. The required theory is that nature is replete with negative feedback devices that correct for human abuses. (Negative feedback is by definition automatic corrective action in a system which restores it to equilibrium when it is disturbed, so usually a good thing.) So when Simon discusses biodiversity, he refers to the fact that nature is always creating species, as well as extinguishing them (Simon and Kahn, 1984: 23). Survivalists reply that the rate of species extinctions caused by environmental destruction far exceeds the speed with which nature can create species; it is the difference between decades and millennia. Wildavsky (1995), in his discussion of global warming, alludes to such feedback devices in ecosystems, but this is a rare admission in the Promethean literature. But if the Promethean position is to stand, such devices must have unbounded capacity to correct for human abuses. In short, what is needed here is an infinitely forgiving nature. Earlier, I noted that Simon referred disparagingly to Santa Claus cornucopians. Yet Promethean discourse requires nature to be more generous still than Santa Claus, who brings only coal and no presents to children who misbehave. For the Prometheans—more properly styled Promethean–Cornucopians—nature will bring good things to us even, and especially, when we misbehave.

Who is right, the survivalists or the Prometheans? Are there limits, or are there not? The survivalist world consists of finite ecosystems with fixed stocks of resources, where human population explosion and economic growth threaten to overshoot the limits of these systems. In the Promethean world, nature does not exist, save as a source of matter to be rearranged in the human interest through application of energy and technology (though an infinitely forgiving nature eventually comes to rescue the otherwise incomplete Promethean worldview). Where Prometheans see benign trends heading off into a happy future, survivalists see looming boundaries into which these trends will eventually crash.

On the trend evidence so far, the Prometheans are clear winners. Now, as we have seen, Prometheans are not always overly scrupulous about a little sleight of hand when it comes to presenting statistics. And given complexity and interdependence in environmental affairs, improvement on one indicator in one place may mask deterioration in another. This is the phenomenon of displacement (see Dryzek, 1987: 16–20). Displacement occurs when, for example, a country exports its toxic wastes, or its polluting industry. A cleaner environment in developed countries, for which we have more and better figures to compose trends, has been purchased in part by transferring manufacturing to developing countries with comparatively lax standards. Displacement across space can also occur when tall smokestacks are constructed on coal-burning power stations to reduce local pollution—only to cause acid rain elsewhere, as a result of sulfur dioxide spending longer in the atmosphere. Displacement can take place across the media, as when a water pollution problem is solved by capturing effluent, drying, and burning it; or disposing of it as toxic sludge. Promethean statistics should be believed only when they refer to global trends. Caution may be warranted there too—if, for example, growing global agricultural production has been purchased at the expense of the long-term productivity of land (through excessive use of fertilizers and pesticides, or farming techniques that hasten soil erosion). This example shows that displacement can occur across time too. Prometheans ignore this issue (or, in Lomborg's case, attempt to refute it by ridiculing one poor example of displacement; 2001*a*: 11). Their confidence in the veracity and power of statistical indicators of environmental trends represents refusal to recognize complexity and uncertainty in ecological affairs.

These issues notwithstanding, when it comes to long-term global trends in natural resource prices, agricultural production, and life expectancy, most of the lines on the graphs have indeed been pointing in benign directions. Thus survivalists are tactically mistaken in taking on Prometheans in arguing about which direction the global trends are pointing—though many survivalists have made the mistake of so doing, ending up with egg on their faces (or, in the case of Paul Ehrlich, a wallet $1,000 lighter). Survivalists in more astute moments will simply argue that the fact that a trend has persisted in the past is no guarantee that it will persist indefinitely into the future. The driver of an accelerating car about to hit a brick wall might well say 'so far, so good'—but that does not mean the wall is not there. Survivalists have given us all kinds of good arguments for the existence of walls—even though they cannot prove that such walls exist, let alone specify precisely how far in front of the car they are located. However, survivalists have done themselves no favors in producing wild scare stories which Prometheans have had fun debunking. False alarms along these lines include predictions of global *cooling* in the 1970s, and claims about imminent exhaustion of oil or particular minerals.

One way to resolve this issue might be to compare the answers to two questions. First, if we believe the Prometheans and they are wrong, what are the consequences? Second, if we believe the survivalists and they are wrong, what are the consequences?

As a footnote to this analysis of Promethean discourse, on dimensions unrelated to limits and survival Prometheans can be found on the environmentalist side. Julian Simon says that he enjoys the outdoors, especially likes birdwatching, and looks back fondly on the nature study merit badge he earned as a boy scout (1996: xxxiv). Aaron Wildavsky concludes his Promethean exposé of environmental risks with the question, 'What, in my vision, is left of environmentalism? There is respect for nature, for all life. There are moral questions of human relationships to all creation' (1995: 447). Given that there is little in Promethean discourse about aesthetics, there is nothing to stop Prometheans being aesthetic environmentalists. Aesthetics aside, if the Prometheans are right then all other discourses of environmental concern are rendered irrelevant and unnecessary.

NOTES

1 For updates, see **www.anti-Lomborg.com**. Lomborg's replies to his critics can be found at **www.Lomborg.com**.

2 *Guardian*, London, January 9, 2003.

3 Thanks to Gerry Mackie for this description.

4 *Observer*, London, April 4, 2004.

PART III

SOLVING ENVIRONMENTAL PROBLEMS

The clash of survivalists and Prometheans detailed in Part II is full of drama, and the stakes appear massive—nothing less than the fate of the Earth. Yet if we look for specific changes in institutions, policies, and practices directly traceable to these discourses, we are often likely to be disappointed. Prometheans would say that the whole point is that nothing much needs changing, though there are a large number of public policy practices they would like to see eliminated, involving fairly radical changes. In practice, we find more limited policy responses in an environmental context. Governments have not engaged in draconian population control or sought an end to economic growth. Instead, they have opened their doors to environmental lobbyists, passed laws to conserve resources or ameliorate pollution, and created bureaucracies to implement these laws.

I turn then to a less apocalyptic discourse that has had obvious consequences in the way societies, and especially governments, have gone about characterizing and attacking environmental problems. The discourses of environmental problem solving recognize ecological problems, but treat them as tractable within the basic framework of the political economy of industrial society. The basic storyline is that of problem solving rather than heroic struggle. Human interactions with the environment generate a range of problems (rather than one big problem like overshoot of limits threatening social collapse), to which human problem-solving devices can be turned. Different varieties of this discourse reveal different conceptions about how best to organize problem solving, especially when social problems require coordination of large numbers of individuals. The three main ways to coordinate such efforts are by bureaucracy, democracy, and markets. Corresponding to these three coordination mechanisms are the three discourses I address in Chapters 4, 5, and 6: administrative rationalism, democratic pragmatism, and economic rationalism. However much partisans of these three variations may disagree with each other, they share the basic storyline of problem solving; and their differences with survivalists, Prometheans, sustainable developers, and green radicals are striking. Of the three, I will deal with administrative rationalism first because it captures the dominant governmental response to the initial onset of environmental crisis. Democratic pragmatism soon emerges as a corrective to administration. And economic rationalism builds on its advances in all areas of political life to generate alternatives to and remedies for the pathologies it identifies in both administration and liberal democratic governance.

4

Leave it to the Experts: Administrative Rationalism

Environmental issues are typically complex. They also involve systems that have long been the objects of study of natural scientists (and public health engineers). Thus when these issues came to prominence in the 1960s they could be associated with a public policy tradition that accorded substantial status to scientific expertise harnessed by administrative structures. This nexus of science, professional administration, and bureaucratic structure has been used in many policy settings: defense and national security planning, public health engineering and health-care delivery, agriculture, and natural resources management. Administrative rationalism may be defined as the problem-solving discourse which emphasizes the role of the expert rather than the citizen or producer/consumer in social problem solving, and which stresses social relationships of hierarchy rather than equality or competition. As an institutional style, administrative rationalism figures more strongly in some political systems than in others. So it is very strong in France and Germany, slightly less strong in Britain (where it has been leavened by a culture of generalism among high-ranking civil servants), and it has had a somewhat patchy but none the less significant presence in the United States.

When environmental issues rose to prominence on the political agenda, their assimilation to this tradition was not planned or debated against the alternatives (such as those that might be generated by the alternative discourses that appear in other chapters). It was simply taken for granted that this was how issues should be handled. Thus the onset of environmental problems was met by institutional and policy responses remarkably similar in both content and timing across the nations of the developed world.[1] If one seeks the essence of most other environmental discourses, it is to be found in the writings and speeches of theorists and activists. But for administrative rationalism, that essence can be captured by looking at actual practice in the development of policies, institutions, and methodologies. Administrative rationalism does have its theorists, but they tended to come later, as they contemplated ways of building upon or strengthening existing environmental accomplishments. Later in this chapter we will encounter some of these theorists, but it is more appropriate to begin with a survey of practices, policies, and institutions.

The repertoire of administrative rationalism

Administrative rationalism manifests itself in the following institutions and practices.

Professional resource-management bureaucracies

'Natural resource management' has been around much longer than 'environmental policy,' especially for governments with resource-rich territory and significant economic activity in the resource sector, notably the United States, Canada, and Australia. The oldest professional resource management bureaucracies are to be found in the United States, the legacy of its Conservation Movement at the beginning of the twentieth century (Hays, 1959). This movement was infused with some German ideas about conservation ecology via its key figure, Gifford Pinchot, who studied in Germany. The movement's main argument was that the American endowment of natural resources was in danger of being squandered in a free-for-all, such that more rational scientific management coupled with government ownership was required to better put those resources to efficient human use. The movement had no interest in wilderness preservation, environmental aesthetics, or pollution reduction, and sought only to achieve maximum sustainable yield from renewable resources such as forests and watersheds. The Conservation Movement achieved ascendancy in Washington DC, and Gifford Pinchot lent his guidance to the administration of President Theodore Roosevelt. The main organizational legacy was the US Forest Service, located within the Department of Agriculture, which was reorganized by Pinchot. However, the Forest

Service's ethos of professional resource management based on scientific principles rather than political expediency could not withstand sustained political pressure from the timber industry. Today, the US Forest Service functions mainly to service the industry—for example, by constructing logging roads into national forests at public cost, so providing enormous public subsidy to the industry. Welfare logging of this sort presumably has Gifford Pinchot turning in his grave.

Later, President Franklin Roosevelt's New Deal in the 1930s saw the establishment of several federal resource management agencies, notably the Civilian Conservation Corps, the Tennessee Valley Authority, and the Soil Conservation Service. Creations of a confident presidency, these agencies were deliberately insulated from congressional influence, thus giving professional managers space without having to worry about political oversight. Ackerman and Hassler (1981: 4–6) define a New Deal agency in terms of the 'affirmation of expertise,' insulated from both political control and judicial oversight. Today, the US federal government is home to resource management bureaucracies such as the Bureau of Land Management, the Fish and Wildlife Service, the National Park Service, the National Oceanographic and Atmospheric Administration, and the US Geological Survey. None of these is a paragon of scientific management to the exclusion of political influence—especially the influence of extractive industry, be it miners, loggers, oil companies, ranchers, or fishers. But all operate according to at least a public justification of administrative rationalism, however much that may be violated in practice. All employ individuals with relevant scientific and professional expertise, many of whom think that rational resource management is what they are doing.

Pollution control agencies

Not every country has a vast national estate of natural resources. But every country suffers from pollution; and so every country that can afford it has a pollution control agency. Many subnational governmental units such as states, provinces, and cities also possess such agencies; they have even emerged at the international level (for example, the United Nations Environment Program, which has brokered pollution control agreements for regional seas).

The oldest such agency is Britain's Alkali Inspectorate, created in 1864. The Alkali Inspectorate is one of the ancestors of the unified Inspectorate of Pollution established in Britain in 1987 as part of the Department of the Environment. This Inspectorate was itself later merged into a still more inclusive Environment Agency. This unification was a bit belated; most developed countries gained such an agency in the early 1970s. So the Netherlands gained a Department for Public Health and Environmental Hygiene in 1971, the US Environmental Protection Agency was established in 1970, and in Germany anti-pollution policy was centralized in the Interior Ministry in 1969, later passing to a free-standing agency. Such agencies are typically charged with implementing laws. Landmark pieces of legislation here include the 1956 Clean Air Act in the United Kingdom (passed in response to London's 'killer fog' in December 1952).

The US Environmental Protection Agency (EPA) is sometimes regarded as the paradigmatic anti-pollution agency, but in fact it is a bit of an anomaly, for its professional discretion is highly constrained. The members of Congress who set up the EPA in 1970 had in mind the experience of regulatory agencies that had been captured by the industries they were supposed to be regulating (so the trucking industry controlled the Interstate Commerce Commission, the food industry controlled the Food and Drug Administration). To prevent such capture, Congress specified in a number of statutes (such as the Clean Air Act, Water Pollution Control Act, Toxic Substances Control Act) in great detail exactly what the EPA must do, setting precise targets and dates for pollution reduction and the means for achieving them. This micro-management on the part of Congress intensified in the 1980s when congressional leaders rightly perceived that the Reagan administration wished to dismantle the EPA and its mission (Rosenbaum, 1995: 208–9). The EPA's counterparts in other countries typically have much more discretion in setting standards and deadlines, and in formulating measures to apply in particular cases.

Resource management bureaucracies and regulatory pollution control agencies rest claims to authority on the scientific and professional expertise they mobilize. Such claims to impartial expertise have come under attack from conservatives and postmodernists, united in a belief that scientific neutrality is impossible, that all science is ideologically colored. The politicization of science in the administration of George W. Bush was criticized in a 2004 report by the Union of Concerned Scientists; *Scientific*

Integrity in Policymaking: An Investigation into the Bush Administration's Misuse of Science. This politicization involved replacing scientists on advisory committees with pro-industry partisans, selective release of the findings of scientific studies of proposed anti-pollution legislation in order to discredit proposals unfavorable to industry, and the suppression of discussions of global warming in EPA publications.

Regulatory policy instruments

Whether as a matter of legislation (as for the US EPA) or of choice on the part of the agency, the most popular policy instrument for pollution control has in all developed countries been regulation (see Opschoor and Vos, 1988 for a survey). Regulation involves the staff of the agency formulating knife-edge standards for particular polluters, who are punished (usually by fines) if and when these standards are not met. Regulators can also specify the kinds of pollution-control equipment that must be installed to clean emissions (for example, catalytic converters on car exhausts, or scrubbers for coal-burning power plants), the kinds of materials that can be used (for example, unleaded gasoline, or low-sulfur coal), and the kinds of practices that must be followed (such as inspections and safety checks). Normally regulation has been 'end of pipe' in character—that is, regulators have not intervened to specify changes in production processes to make them produce less noxious waste. Instead, the focus is on reducing discharge of that waste into the environment once the waste has been produced.

Regulation of this sort entails substantial discretion on the part of the regulators, even in the United States. That said, national approaches to regulation vary. In the United States, regulation proceeds in adversarial fashion, and both sides rely a great deal on lawyers to advance cases for more or less stringent pollution standards. Many decisions end up in the courts: polluters sue the EPA for excessive stringency in enforcing the relevant law and for arbitrary action violating the principles of the US constitution; environmentalists sue the EPA for not enforcing laws with enough vigor; the EPA sues polluters for not complying; corporations sue individuals for defamatory criticisms (so-called SLAPP suits—strategic lawsuits against public participation). Thus in the United States it is the courts' interpretation of legislation that is decisive, and administrative rationalism in practice is highly constrained by this legalistic, adversarial context.

Matters are very different in countries where administrative rationalism is allowed a freer hand. In Britain, where administrative rationalism has traditionally dominated environmental policy (Gray, 2000), regulations are developed in consultation between government officials and polluters. Pollution abatement and environmental quality appear no worse in Britain than in the United States (see Vogel, 1986). Regulation in Britain is not simply a matter of political negotiation, for all sides at least in principle accept the authority of scientific expertise in adjudicating disputes. In some cases, a scientific body is called on formally to render a verdict—the Royal Society played this role on acid rain policy in Britain in the 1980s, effectively arbitrating disputes (Hajer, 1995: 144–5). Pollution-control discourse in Britain specifies that no regulatory action be taken until science can demonstrate the harm being caused. As William Waldegrave, Minister of State in the Department of the Environment, put it in 1987: 'It is necessary in an area which should be science-based to put up pretty formidable hurdles and tests of a scientific nature if we are to make rational priorities' (quoted in Weale, 1992: 80). This is the exact opposite of the 'precautionary principle' applied in countries such as Germany and the Netherlands, which specifies that scientific uncertainty is not a good reason for delaying action (see Chapter 8).

Environmental impact assessment

Environmental impact assessment specifies that government departments (and, in some cases, private developers) must prepare a systematic assessment of the environmental damage likely to be caused by any project, be it an airport, a mine, a shopping mall, a sale of oil or mining or timber leases, a freeway, or a pipeline. Typically, only projects with anticipated major environmental impacts are covered. The intent is to force consideration of environmental values and scientific means for the calculation of the project's effect on them. The US National Environmental Policy Act of 1970 (NEPA) is a landmark in establishing this kind of process, but again the US turns out to be slightly anomalous. Large numbers of impact statements have been prepared in the US, perhaps the most famous being the two prepared for the Trans-Alaska Pipeline immediately following the passage of NEPA. The first was a few pages long; after being ruled inadequate by the courts, a second was prepared in multiple volumes which would occupy many feet of shelf space (ironically, the pressure of the energy crisis in the wake of the 1973 Organization of Petroleum Exporting Countries oil embargo eventually led Congress to exempt the pipeline from NEPA requirements). Again, the courts have played a large role in determining what US legislation actually means. In the case of NEPA, the courts' interpretation has been that an impact statement must be prepared, not that it must actually be used in decision making. Thus US environmental impact statements became long and unreadable documents, designed to defend agencies against accusations that environmental concerns were not being taken seriously. Environmentalist and community objectors to proposals could file suit only on the basis of the adequacy of the impact statement, not on the basis of the substantive merits of the agency's decision. Observers of the environmental impact process in the US are divided on whether it has indeed improved policy making as intended. Even Lynton Caldwell, largely responsible for crafting the NEPA legislation, could not make up his mind on whether the Act had the desired impact (for a pessimistic assessment, see Caldwell, 1978; for a more positive verdict, see Caldwell, 1982).

Environmental impact assessment soon spread to Canada, Australia, Germany, France, and elsewhere. Again, Britain dragged its feet, and only accepted environmental impact assessment at the behest of the European Community in 1985. In these other countries the path of administrative rationalism in environmental impact assessment has been less strewn with legal and political obstacles than in the United States. Not that matters are always smooth. For example, in Melbourne (Australia) the state government (Victoria) routinely exempted the biggest and most controversial projects from assessment. In the mid-1990s these included construction of a Grand Prix motor-racing circuit in an inner-city park, a freeway network around the city center, and the world's biggest casino.

Environmental impact assessment, even when it can escape legal shackles, is not unalloyed administrative rationalism. For typically the process also mandates opportunities for public comment on impact statements and so public participation in the policy process. This latter aspect of impact assessment is more easily joined to democratic pragmatism, and will be discussed in the next chapter.

Expert advisory commissions

The United States also pioneered the idea of an expert commission to offer advice on environmental affairs. The President's Council on Environmental Quality (CEQ) was set up in 1970 in a section of the National Environmental Policy Act. Quite what the commission would do (aside from offer comment on the environmental impact processes established under NEPA) was never entirely clear, and the role of the CEQ has varied substantially across different presidencies. It fell into virtual disuse in the 1980s. The CEQ might act as a counterweight to the longer-established and more influential Council of Economic Advisors. Both report directly to the President. But the CEQ has never attained the standing of the CEA. In 1993 President Clinton merged the CEQ into an Office of Environmental Policy (OEP) in the White House, a development which had the appearance, though not necessarily the substance, of an upgrading. This arrangement continued under President George W. Bush, though the CEQ/OEP profile continued to fade. It remains ironic that bodies such as Germany's Council of Environmental Experts, established in 1972 and modeled on the US CEQ, have achieved greater centrality in policy making.

In Britain, there is a long tradition of deference to scientific expertise, and so to expert advisory bodies. Notable among such bodies is the Royal Commission on Environmental Pollution, established in 1971. Its mandate is narrower than that of the US CEQ; but its policy role is greater. In addition, the Royal Society is occasionally called upon to play a key role in policy determination. None of this means that environmental policy in Britain is made by administrative rationalism untarnished by politics, for science is expected to fall into line with the policy priorities of the government of the day. Given that one of those priorities is normally that more scientific research is needed before any substantial action is taken to protect the environment, it is not hard to find scientists who can tailor their recommendations accordingly.

Rationalistic policy analysis techniques

The expertise that legitimates administrative rationalism comes largely in the form of environmental science and engineering. Relevant disciplines include forestry, oceanography, meteorology, ecology, hydrology, geology, fisheries biology, biochemistry, and toxicology. But administrative rationalism also involves application of general-purpose policy analysis techniques, most of them geared to identification of the optimal policy in a given situation. Many of these techniques were developed in an environmental context. The most widely used are cost–benefit analysis and risk analysis. Computer modeling of the sort discussed in Chapter 2 on survivalism is also congenial to administrative rationalism, but less widely used. Other available techniques include technology assessment, decision analysis, and a range of forecasting methods. Cost–benefit analysis can be either forward-looking, in informing the choice of policy or project, or backward-looking, to evaluate policies already in place. Forward-looking cost–benefit analysis involves the following steps:

1. Identify policy options. (The procedure still works if there is only one option to compare with 'do nothing.')
2. For each option, list both desirable effects (benefits) and undesirable effects (costs).
3. Attach monetary values to all costs and benefits, using 'shadow pricing' when the item in question has no market price.
4. Convert all costs and benefits occurring in future time periods to the present time period using a discount rate.[2]
5. Add up the monetary costs and benefits to give the net benefit associated with each alternative.
6. Choose the option with the greatest net benefit (provided that this net benefit is positive).

The real substance of a piece of cost–benefit analysis is the shadow pricing. Obviously some items can easily have a monetary cost pinned onto them. For example, if the analysis is of a proposal to build a dam, construction costs can easily be expressed in terms of dollars. So can the benefits of the electricity generated by the dam. Other items are more difficult. How does one value in monetary terms the loss of a free-flowing river? Or the benefits to recreational users yielded by the artificial lake that will be constructed? Many shadow-pricing techniques can be brought to bear. A lost environment (such as a drowned river valley) can be valued by conducting a survey and asking individuals how much compensation they would require to consider themselves no worse off than before. Alternatively, they can be asked how much they would be willing to pay to prevent the drowning. Or the amount of time and money individuals expend to reach the valley for recreational activities can be observed and summed and used to calculate how much individuals actually pay to get to the valley. Valuing lives saved is still more controversial. In a cost–benefit analysis conducted by the US EPA in connection with the George W. Bush administration's Clear Skies initiative in 2003, a 'senior discount' was applied that reduced the value of the lives of old people, leading to a political storm (Schmidt, 2003).

Cost–benefit analysis was pioneered in the siting and construction of dams in the United States, starting in the 1950s. The main sponsors were the US Army Corps of Engineers and the Bureau of Reclamation. One of the more famous pieces of cost–benefit analysis was conducted in the late 1960s in Britain by the Roskill Commission, set up to recommend a site for a third major airport for London. Roskill's recommendation in favor of a site at Wing in Buckinghamshire was reached via a cost–benefit analysis of the alternative sites. In its efforts to monetize all costs and benefits associated with each site, the Commission provided plenty of ammunition for its opponents. For example, the price put on a centuries-old church which would have to be demolished was determined by the increased travel time churchgoers would have to spend to go to more distant churches. (The recommendation in favor of Wing was not accepted by the government.)

Cost–benefit analysis received a major boost with the promulgation in 1981 of Executive Order 12291 by President Reagan, which specified that all significant federal regulations, including environmental ones, had to pass a cost–benefit test administered by the Office of Management and Budget. This order owed more to right-wing ideology than to administrative rationalism. The idea was to use cost–benefit analysis as a tool in the Reagan administration's attack on positive government, to free corporations from regulations that harmed profitability. The Clinton administration reaffirmed the application of cost-benefit analysis to regulation. Under the George W. Bush administration, cost-benefit analysis was forced on the EPA and other agencies by the Office of Information and Regulatory Affairs within the Office of Management and Budget (Schmidt, 2003).

Cost–benefit analysis is the subject of a huge literature, both technical in terms of how to do it (see, for example, Sugden and Williams, 1978), and critical in terms of why it should never be done (for example, Bobrow and Dryzek, 1987: 27–43; Sagoff, 1988). From the discourse analyst's perspective, the main cumulative impact of cost–benefit analysis may be in legitimating the idea that public policy is a matter for technical, expert choice and not a question on which non-specialists such as elected officials, still less any broader public, have any rightful say. And this is why cost–benefit analysis rests more easily in a discourse of administrative rationalism than it does in economic rationalism. For this technique uses markets only to provide prices for the balance sheet of costs and benefits. Once such prices have been input, expert-guided governmental actions are central. Thus cost–benefit analysis has an implicit faith in the welfare-maximizing virtues of government officials, which true economic rationalism lacks, preferring instead that market mechanisms be utilized wherever possible.

Risk analysis covers a family of procedures and techniques, to quantify the potential harm from environmental hazards, such as ingesting pollutants, living downwind of a nuclear power plant, or being exposed to additional ultra-violet radiation as a result of ozone depletion in the stratosphere. It is sometimes possible to compute a dose–response curve, which shows how the risk to health and life varies with exposure to a hazard. The main sources of information in risk assessment are animal studies and epidemiology. Animal studies are based on the assumption that exposing animals to high doses of a pollutant can yield useful information about what happens when human beings are exposed to much lower doses (in terms of cancer rates, etc.). Epidemiological studies are statistical analyses of human populations that relate degree of exposure to a risk (for example, quantity of suspended particulates in the air) to the incidence of particular kinds of death and disease (for example, lung cancer). Both instruments can be very blunt. As Wildavsky (1995: 254, quoting David Ozonoff) puts it, 'a good working definition of a catastrophe is an effect so large that even an epidemiological study can detect it.' Thus it is no surprise when the claims of activists about health damage from a toxic dump in the neighborhood, electromagnetic radiation from power lines, or herbicide spraying are not confirmed. The number of cases is usually too small to enable demonstration of population-level effects required for statistical significance in an epidemiological study; but this does not mean the activists are wrong (Tesh, 2000). O'Brien (2000) suggests that a focus on risk assessment means ignoring alternatives; for example, examining the risks associated with a toxic waste incinerator can in practice obscure the question of why toxic wastes should be generated to begin with. Also, risk assessment is not very good at dealing with the interaction effects of multiple environmental hazards.

Risk analysis also involves study of perceptions by ordinary people. Almost invariably, the scientific evidence shows that people wildly overestimate the potential damage to them from environmental risks such as having a toxic waste dump in the vicinity (Wildavsky, 1995), and behave quite inconsistently in relation to risks. Given that both animal studies and epidemiology are extremely blunt instruments, public skepticism and alarmism become a bit more understandable. Still, thinking of the place of risk analysis in environmental discourse, its political function is quite clear: the assessment of true risks is a matter for the experts, and the ordinary public usually gets it wrong.

Many of the psychologists who study risk give greater credence to public skepticism (for example, Fischoff et al., 1982). Public skepticism can be explained by, for example, distinguishing between risks that are borne voluntarily, like driving a car, and those that are incurred involuntarily, like being exposed to fumes from a nearby factory. People have a much higher tolerance for risks they bear voluntarily. They also tend to have a low tolerance for low-probability catastrophic events (such as the meltdown of a nuclear reactor), and so overestimate the risks from them.

Discourse analysis of administrative rationalism

Administrative rationalism seeks to organize scientific and technical expertise into bureaucratic hierarchy in the service of the state. As such, it rests on the following components.

Basic entities whose existence is recognized or constructed

Administrative rationalism is a problem-solving discourse, and so takes the structural status quo of liberal capitalism as given. Within this status quo, the discourse has a strong conception of the nature of government.

Government is the administrative state, treated in monolithic terms. Governing is therefore not about democracy, but about rational management in the service of a clearly defined public interest, informed by the best available expertise. Managers and experts have well-defined roles within the administrative monolith.

The best real-world approximation to this monolithic imagery could long be found in Germany, home to a 'legal corporatist' conception of government under which law is seen as an expression of state authority. In environmental administration, law was to be based on the best science. The Prussian administrative tradition views state and society as an organic whole, and the public interest in abstract legal terms, so not something that interest groups should bend. Opponents of the implementation of law could be styled as obstructionists and cranks, even if they were denied input to the formulation of law. While this system of government began to break down in the 1980s, residues linger in the new century (Dryzek et al., 2003: 35–42).

Administrative rationalism is quite agnostic about many of the entities that so energize survivalism and its Promethean critics—entities such as ecosystems, finite stocks of resources, population, energy (at least in the key role Prometheans construct for it). There is nothing to stop administrative rationalism coming across such concepts in the problem-solving quest; but they are not at center stage.

Assumptions about natural relationships

While not explicitly concerned with the fundamental character of relationships between human and nonhuman worlds, administrative rationalism does assume that nature is rightfully subordinated to human problem solving, though not in the forthright and confident celebration of human domination of nature found among Prometheans.

Within human society, administrative rationalism assumes two complementary kinds of hierarchy. The first subordinates the people to the state. The second puts experts and managers in their properly dominant places in the state's own hierarchy, which is justified on the basis of expertise. The discourse pretty much denies the existence of politics of any sort.

Agents and their motives

Agency is granted to both collective and individual actors. Government as a collective actor is the primary agent, but this does not imply that all individuals working for the state have an equal capacity to act. Technical experts and managers have a greater capacity than anyone else. Motivations are treated as entirely public-spirited, and the public interest is conceptualized in unitary terms. Thus discovery and application of the public interest is itself a technical procedure (see Williams and Matheny, 1995: 11–17), which is why (for example) cost–benefit analysts or risk assessors know better than the public itself what is in the public interest.

Key metaphors and other rhetorical devices

Administrative rationalism is much less vivid in its metaphors than survivalist and Promethean discourses. Doom and redemption are not at issue, which makes for muted rhetoric. Environmental problems are serious enough to warrant attention, but not serious enough to demand fundamental changes in the way society is organized. Thus the rhetoric combines a mixture of concern and reassurance, both of which can be drawn upon at particular stages in problem-solving efforts. So government officials can reassure people that there is no cause for alarm when a particular environmental risk surfaces (be it asbestos in schools, radon in basements, or genetically modified organisms in the environment). The same actors may also point out that policy measures need to be taken, though normally in ways that treat the risk in piecemeal fashion, rather than as a manifestation of anything more deeply wrong with industrial society.

If there is a metaphor that characterizes the discourse, it is that of a unitary and omniscient administrative mind. This is like the human mind, only collective and embodied in the administrative state. Just as the human mind controls the body, so the administrative mind controls the state. As Torgerson (1990: 120–1) puts it: 'The image of the administrative mind is one of an impartial reason exercising unquestionable authority for universal wellbeing; it is an image which projects an aura of certain knowledge and benign power.'

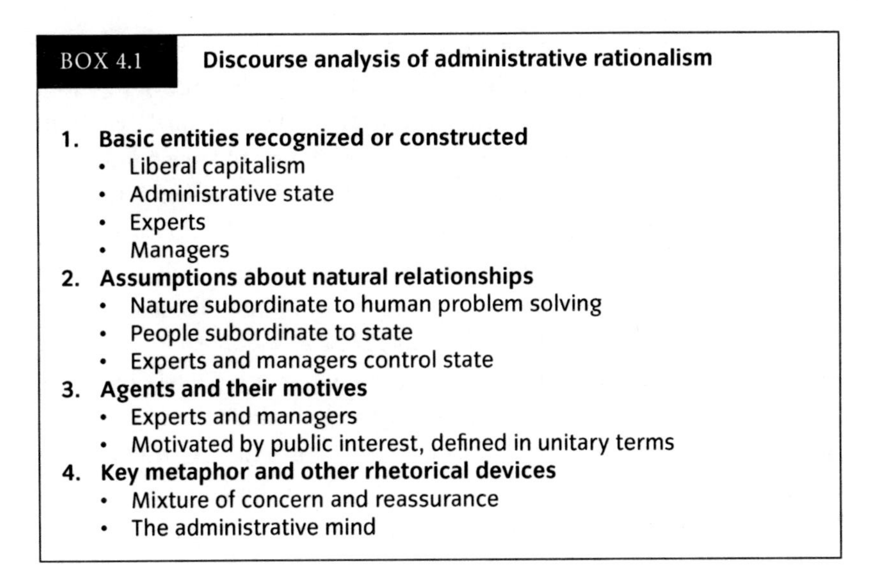

BOX 4.1 **Discourse analysis of administrative rationalism**

1. **Basic entities recognized or constructed**
   - Liberal capitalism
   - Administrative state
   - Experts
   - Managers
2. **Assumptions about natural relationships**
   - Nature subordinate to human problem solving
   - People subordinate to state
   - Experts and managers control state
3. **Agents and their motives**
   - Experts and managers
   - Motivated by public interest, defined in unitary terms
4. **Key metaphor and other rhetorical devices**
   - Mixture of concern and reassurance
   - The administrative mind

The justification of administrative rationalism

I argued earlier that the search for administrative rationalism should begin not with the writings of theorists and the proclamations of activists, but with an examination of actual policy practice. I have defined that practice in terms of the sixfold repertoire of administrative rationalism. These items are for the most part institutional and policy hardware, with very tangible existence. As bits of hardware, some of them can be appropriated by competing discourses, at least by the other problem-solving discourses set out in the next two chapters. The essence of administrative rationalism is to be found in the discursive 'software' that unites these six items around a common purpose. As a problem-solving discourse, administrative rationalism takes the political-economic status quo of liberal capitalism as given. It then puts scientific and technical expertise, organized into bureaucratic hierarchy, motivated by the public interest, to use in solving environmental problems without changing the structural status quo. With this characterization in hand, it is possible to identify more clearly the justification on which administrative rationalism rests.

The twentieth century was greeted by the German sociologist Max Weber with an announcement that bureaucracy was the supremely rational form of social organization (see Gerth and Mills, 1948). Weber was not happy about this, but saw increasing rationalization of society through bureaucratic organization as inevitable. Increasing complexity in social and economic problems could not be confronted by individuals acting in isolation, only by coordinated problem-solving efforts of large numbers of individuals. The best way to cope with a large, complex problem is to break it down into smaller sets and then into still smaller subsets. Each subset should then be assigned to an individual or small group to craft a solution. These partial solutions are then aggregated into a solution for the complex problem as a whole (see H. Simon, 1981). Clearly, somebody needs to formulate the initial breakdown into sets and subsets; somebody needs to keep an eye on the people dealing with each subset; and somebody needs to piece together the elements. That 'somebody' is the apex of organizational hierarchy. Hierarchy is justified on the basis of access to both principles of administrative management and substantive expertise in the issue area in question. The structure of problem disaggregation and solution aggregation describes too the standard organization chart of a bureaucracy.

When Weber was writing, environmental problems were not conceptualized as such. But when these problems did reach the agenda, Weberian bureaucracies were constructed to deal with them. How is an anti-pollution agency organized? It is normally divided into offices dealing with air pollution, water pollution, hazardous wastes, and solid wastes. Each of these offices is then disaggregated further, perhaps on a regional basis, perhaps to deal with different kinds of pollutants, perhaps to deal with different industries. For example, the air pollution office could be divided into 'stationary sources' (smokestacks) and 'mobile sources' (vehicles). The stationary source unit could then be divided further into units dealing with power generation and manufacturing. Power generation could be divided into suspended particulates (smoke), greenhouse emissions (carbon dioxide), and sulfur dioxide (acid rain). Particular kinds of scientific and engineering expertise can then be assigned to the appropriate unit or sub-unit.

In practice, political factors often intervene to confuse the organizational chart. These factors include interventions by elected officials, political parties, lobbyists, and occasionally even the public. When this happens, policy making can become more complicated and messy. The response by structural reformers is often to try to depoliticize and centralize decision making. For that especially politicized agency, the US EPA, Walter Rosenbaum (1985: 299–300) recommends four measures along these lines: a fixed five- or seven-year term for the Administrator (head of the EPA) so that he or she cannot be dismissed at will by the President; the replacement of political appointees by professional civil servants; the establishment of an inspector-general to oversee the professional conduct of the EPA and its employees; and an external scientific review of the EPA's use of technical information in decision making.

This sort of insulation is also recommended by Bruce Ackerman and William Hassler (1981) after their analysis of the disastrous consequences of subjecting the US EPA to political micro-management by Congress. The consequence of that micro-management is that battles are fought in the congressional arena over the means of policy—and questions of means, they argue, are surely a matter for the experts, not politicians and lobbyists. The particular case they analyze is the Clean Air Act Amendments of 1977, which mandated installation of expensive scrubbers to clean the emissions of all new coal-burning power stations. This proposal was supported by both high-sulfur coal producers from the eastern states and western-based environmentalists concerned only with pristine air and minimal intrusion into wilderness in the western states. The proposal was both hugely expensive and likely to lead to an increase in the quantity of acid rain, because it discriminated against low-sulfur coal mined in the West. Surely no expert agency would ever have produced such a perverse policy. The solution to the EPA's woes, they argue, is to have Congress force very precise ends on the agency—expressed, for example, in person-years of life expectancy added to the US population by a particular date (p. 124). This precise specification will protect the agency against industry capture, the fear of which was the reason for congressional micro-management of the agency to begin with. Once these ends have been imposed, EPA professionals should be left alone to craft appropriate solutions.

If political conflict cannot be banished from environmental administration, the next best thing for the administrative rationalist may be to channel it more productively. Kai Lee (1993) argues that the proper function of political conflict is to raise issues which managers might otherwise miss. Lee's ideal is 'ecosystem management,' in which professional resource managers trained in ecology take charge of whole ecosystems. He regards the Northwest Power Planning Council, responsible for the

Columbia River Basin (and on which he served) as exemplary, though it was not in fact governed by ecologists. For Lee, ecosystem management cannot be entrusted to politicians, who do not have the patience to learn, the willingness to tolerate failure for the sake of learning, or the ability to operate on a biological rather than an electoral timescale, or to look beyond the short-term interests of their constituencies. But recognizing that political conflict cannot be eliminated, he suggests taming it by establishing alternative dispute-resolution forums where people can reason through their differences, rather than waste energy in adversarial processes that produce only stalemate or uncreative either/or decisions. Unfortunately, ecosystem management in the Columbia basin ended in the face of legal challenges from environmentalists under the Endangered Species Act.

Administrative rationalism in crisis

Among those who have reflected upon administrative rationalism, increasingly few have done so in order to defend it. Part of this is due to the association with bureaucracy. It is hard to find anyone who actually likes bureaucracy, which is generally defended as necessary rather than attractive. Still, a discourse can soldier on without reflective defenders—indeed, particular discourses may persist precisely because nobody at all is reflecting on them. Unfortunately for administrative rationalism, it is meeting with reflection, much of it very critical.

Prosaic and uninspirational though it might be, administrative rationalism could always sustain itself so long as it delivered the goods. These goods would include cleaner air and waters, fewer toxins circulating in the human environment, an environmentally secure future, improving aesthetic standards in city, suburb, countryside, and wilderness, more securely protected ecosystems and species. But the administrative state's performance on these standards has been called into question. This questioning can often be put under the heading of 'implementation deficit'—a substantial gap between what legislation and high-level executive decisions declare will be achieved and what is actually achieved at street level (Weale, 1992: 17–18). Implementation deficit was originally a German expression. Germany was long the exemplar of administrative rationalism in environmental policy, but it only became an environmental leader once its administrative system began to open up in the 1980s (Dryzek et al., 2003).

The administrative state may be running out of steam in the environmental arena, or experiencing diminishing returns to effort. This would accord with experience in other policy areas such as crime, public health, industrial development, and education. It is relatively easy to achieve substantial initial gains, because the easier and most visible problems will be attacked first. It is very hard to show sustained improvement on any dimension once initial gains have been made. For example, in air pollution control it tends to be suspended particulates in cities that get attacked first: these are (literally) visible and easily remedied by technical fixes (such as mandating smokeless fuels). More insidious pollutants such as lead from car exhausts take longer to come to attention and receive their fix—but eventually they get it in the form of unleaded gasoline. More complex, invisible, and contentious issues like acid rain eventually come to the fore, but prove much harder to conceptualize, to even define the problem at hand, and to craft solutions. As Lindblom (1977) puts it, centrally administered systems have 'strong thumbs, no fingers.'

What lies at the root of these problems? To begin, administrative rationalism implies hierarchy based on expertise, with both power and knowledge centralized at the apex. Those at the apex are assumed to know better than subordinate levels, so as to be able to assign tasks and coordinate operations. But problems of any complexity defy such centralization: nobody can possibly know enough about the various dimensions of an issue such as acid rain, global climate change, ozone depletion, or the interacting cocktail of urban air pollutants, not to mention the social and economic aspects of these issues, to sit with any confidence at the apex. As the philosophers Karl Popper (1966) and F. A. von Hayek (1979) argued at length (though never in the context of environmental problems), the relevant knowledge is dispersed and fragmentary. The closed, hierarchical style of administrative rationalism has no way to aggregate these pieces of information in intelligent fashion. Popper's solution is the give and take of liberal democracy; Hayek's is the market. Popper's critique is especially devastating because it is rooted in a model of science as the exemplary problem-solving activity. To Popper, the hallmark of the scientific community is not authority based on expertise, but free, open, and equal criticism and test of the conjectures of scientists by other scientists. Just as hierarchy and deference to expertise can only obstruct scientific problem solving, so they can only obstruct problem solving in policy and politics.

The Weberian approach to problem disaggregation and assignment of chunks of the problem to different units within the organization requires that such disaggregation be done in intelligent fashion. The main principles here are that interactions within problem subsets as defined should be rich, and interactions across different subsets should be weak (Alexander, 1964). But for truly complex problems, those with a large number and variety of elements and interactions facing a decision system, no intelligent disaggregation may be possible. For defensible disaggregation requires minimization of interactions across the problem chunks. High orders of complexity mean that such interactions will always occur, no matter how intelligently the disaggregation is done. When that happens, there is little problem solving, but a great deal in the way of problem displacement (see Dryzek, 1987: 16–20, 99–100). Such displacement occurs when an air pollution problem is solved by creating a water pollution problem—for example, prohibition of the burning of waste may lead a company to discharge the waste in watercourses instead. It occurs when tall smokestacks are built to alleviate air pollution in the vicinity of a factory, thus leading emissions to fall somewhere else and, in the case of coal-burning power plants, for sulfur dioxide to stay in the atmosphere long enough to constitute acid rain. Most anti-pollution agencies operate under single-medium statutes such as clean air and clean water acts, which facilitate displacement across the media.

This question of displacement has been recognized by administrative rationalists, but rarely answered effectively. The US EPA has experimented with a 'cluster' approach to pollutants, to coordinate rules coming from different pieces of legislation for (say) an industry (Fiorino, 1995: 210). The EPA has also tried 'integrated environmental management' of particular geographical pollution hot spots (see Mosher, 1983). But these efforts have been unable to transcend the divisions across the single-medium statutes under which the EPA continues to operate.

In Britain, the unified Inspectorate of Pollution was created in part to integrate anti-pollution efforts, and so help identify courses of action that would best reduce environmental damage for any given cost to society. But this new Inspectorate was still composed of single-medium units, dealing with industrial air pollution, 'wastes,' radioactivity, and water quality. For a number of years its water operations faced competition from the single-medium National Rivers Authority, though eventually the competitors were merged into the Environment Agency. Legal changes that would have allowed the Inspectorate and its successor to try to operate in holistic fashion were very slow in coming. Thus the integration that has been achieved in the UK is limited (Weale, 1992: 104–7).

To date, the most effective integration has been achieved in Sweden, initially at the plant level (Weale, 1992: 98–9). Later, Sweden pioneered integration across areas with large implications for the environment, but traditionally dealt with separately, by creating a Delegation for Ecologically Sustainable Development made up of the ministers responsible for agriculture, environment, education, labor, and taxation. But such integration remains an uphill struggle in traditionally compartmentalized governments (Lafferty and Hovden, 2003).

A more straightforward reason that helps to explain implementation deficit under administrative rationalism is the problem of compliance with policy decisions. Compliance is required in two stages: first, 'street level' agency officials must comply with legislative direction and the desires of their superiors, and second, polluters, developers, and resource users must comply with directives emanating from the administrative structure. Both kinds of compliance are problematical. As Weale (1992: 18–19) notes, the latter kind of compliance is rarely a matter of actors being told what to do by public officials and then doing it; more often, compliance is negotiated. So, for example, the degree of pollution reduction, the timetable at which it is achieved, and the kinds of equipment to be installed can all be a matter for negotiation. Negotiated compliance makes sense for street-level bureaucrats because they need to cultivate working understandings with polluters, and there is every reason for them to make discretionary judgments about who is to blame and what negotiation strategy is likely to pay off in the long run. Policies made centrally are rarely sensitive to the local circumstances in which street-level bureaucrats operate.

Learning about what works in practice is most easily achieved at this street level. Administrative learning in an environmental context has been pursued in the form of 'adaptive management' of ecosystems, requiring some ecosystem-specific authority to begin with (there are connections to bioregionalism; see Chapter 10). Ideally, this authority would proceed under 'learning by doing,' modeled on scientific experimentation and recognizing high levels of uncertainty. Adaptive management has found its way into the rhetoric of the US Forest Service, among other agencies. However, it has currently seen little in the way of implementation (Lee, 1999). While attractive to scientists, it sidelines both politicians impatient with the long time scale adaptive management requires, and also administrators uneasy with explicit recognition of uncertainty.

The structure of administration often prevents learning being communicated up the administrative hierarchy. As one ascends the administrative hierarchy, the limited time and information-processing resources of individual administrators means that much information is inevitably lost. So administrative rationalism is faced with a conundrum: the more an organization is disciplined, the less it can be expected to learn (see Torgerson and Paehlke, 1990: 9–10). The more it learns (by developing an open and decentralized structure), the less easy will it be to maintain discipline by administrative means, and so the more likely becomes implementation deficit.

From government to governance?

Can administration both learn locally and remain coordinated by adherence to central goals and principles? Sabel et al. (2000) suggest this can be done through what they call a 'rolling rule regime' kind of regulation, under which a central agency sets standards, while compliance is negotiated locally with activists and corporations. The central agency holds the locals to account, but is willing to adjust standards in light of feedback. Sabel et al. point to US cases where this approach has worked, such as watershed protection in Chesapeake Bay and management of toxic pollutants in Massachusetts. This approach is consistent with the principles of 'national standards, neighborhood solutions' and 'collaboration, not polarization' endorsed by federal EPA Administrator Mike Leavitt in 2004.[3] Administrative rationalism is diluted. Skeptics would say there is also an abdication of public authority in favor of unrepresentative private interests powerful enough to secure a place in the dialogue, raising the old specter of capture by industry.

The rolling rule regime approach is also consistent with widespread enthusiasm for 'governance' as opposed to 'government' in public policy in general. Government here is seen as top-down and Weberian, while governance is decentralized, informal, and networked. Governance in this sense is more easily linked to democratic pragmatism than to administrative rationalism, and so will be addressed in the next chapter.

Any wholesale transition from government to governance would signal the end of administrative rationalism. But there may be life in the beast yet. Vogel (2003) argues that in the United States there is actually movement toward greater use of cost–benefit analysis, risk assessment, and more technocratic policy-making in general, accompanied by greater trust in regulators. He believes this return to administrative rationalism results from reactions against regulatory overkill associated with public alarmism about risks. The Superfund clean-up of toxic sites and programs to remove asbestos from schools were hugely expensive and may even have increased risks (by moving hazardous materials that were safely in place). In 1996 Congress modified the Delaney Clause that had long prohibited any substance associated with any risk of cancer in food, in favor of a risk–benefit approach. Vogel notes that the European Union is moving in exactly the opposite direction, as scandals such as mad cow disease (BSE) led to public distrust of government regulators and demands for greater public participation in policy making. Vogel's argument on the United States does not quite add up, because it implies that when regulatory overkill goes, technocracy returns. But overkill itself was often administered technocratically. Thus his conjecture about the return of administrative rationalism should be treated with caution. All it suggests is that some aspects of administrative rationalism are tenacious in the face of 'new governance' trends.

Given the locally variable and negotiated ways in which administrators now secure compliance, it is ironic that market-oriented opponents of administrative rationalism still try to stigmatize it by attaching the ugly and undemocratic term 'command and control' to regulatory policy. This is a rhetorical accomplishment, but very little commanding and controlling actually goes on in the implementation of environmental regulations.

For better or for worse, administrative rationalism has had substantial impact in the environmental arena. Today it may be running out of steam and facing crisis, perhaps even yielding to governance, but its past achievements should not be forgotten. The countries of the developed world have an environment which is cleaner, safer, and more aesthetically pleasing than it would have been without the last forty years of administrative rationalism. This evaluation does not mean that administrative rationalism was the most effective conceivable response to environmental crisis, or even an adequate one. Nor are past achievements any guarantee of future success. So let me turn to the two other problem-solving discourses, which are presented by their adherents as containing effective remedies for the contemporary ills of administrative rationalism.

NOTES

1 For the United States, Hays (1987: 393–4) argues that the high-visibility conflictual environmental politics of the late 1960s and early 1970s yielded to environmental administration and management in the late 1970s, a development he describes as the rise of the environmental professional. In contrast, I would argue that the administration was there all the time in this era, but that in the late 1970s the politics became fairly subdued, leaving the field relatively free for the professionals. Political conflict has certainly made many comebacks since then.

2 A discount rate is like an interest rate, except that it works back from the future to the present. So at a discount rate of 5%, $100 expected one year in the future has a present value of $95. Choice of a discount rate can be quite controversial, and different discount rates can impart different conclusions to the analysis.

3 http://**www.epa.gov/adminweb/leavitt/enlibra.htm**

5

Leave it to the People: Democratic Pragmatism

Ours is a democratic age; it is increasingly unfashionable for anyone, anywhere in the world, to proclaim themselves to be anything but a democrat. It is correspondingly hard to proclaim one's faith in bureaucracy and administrative rationalism. Administration is not very popular as an ideal; rather, it is just what a lot of people, and a lot of institutions, actually end up doing—even the people doing it rarely admit to liking it. Democracy is different; everyone wants to be a democrat. Whether they truly are democrats is a different question, made harder to answer by the sheer variety of meanings and models of democracy.

In this chapter I treat democracy not as a set of institutions (elections, parliaments, parties, etc.), but rather as a way of approaching problems. I will be concerned with democracy as a problem-solving discourse reconciled to the basic status quo of liberal capitalism. This is the version of democracy which dominates today's world, especially after the revolutions of 1989 destroyed the credibility of some Marxist alternatives. Other discourses of democracy do exist, some of which challenge this status quo, advocating radical participatory alternatives, for example (see Dryzek, 1996*a*), and I shall return to some of these alternatives in later chapters.

Democratic pragmatism may be characterized in terms of interactive problem solving within the basic institutional structure of liberal capitalist democracy. The word 'pragmatism' can have two connotations here, both of which I intend. The first is the way the word is used in everyday language, as signifying a practical, realistic orientation to the world, the opposite of starry-eyed idealism. The second refers to a school of thought in philosophy, associated with names such as William James, Charles Peirce, and John Dewey. To pragmatist philosophers, life is mostly about solving problems in a world full of uncertainty. The most rational approach to problem solving, in life as in science, involves learning through experimentation. For problems of any degree of complexity, the relevant knowledge cannot be centralized in the hands of any individual or any administrative state structure. Thus problem solving should be a flexible process involving many voices, and cooperation across a plurality of perspectives. The degree of participation with which pragmatists are happy often corresponds roughly to that found in existing liberal democracies (though some pragmatists advocate more and better participation), so there is an essential congruence between the demands of rationality in social problem solving and democratic values.

Pragmatist philosophy received an explicit environmental twist with the arrival of 'environmental pragmatism,' which takes its bearings from philosophical debates in the field of environmental ethics. Environmental pragmatism does battle with all attempts to propose moral absolutes to guide environmental affairs, which are treated instead as ripe for tentative problem-solving efforts in which a plurality of moral perspectives is always relevant (see Light and Katz, 1996). In this chapter I will be concerned less with the finer points of environmental pragmatist philosophy, and more with the way democratic and pragmatic discourse plays out in the real world of environmental affairs. This style of problem solving can be found within administrative structures, in negotiations between parties to a legal dispute, in international negotiations, in informal networks, as well as in legislatures.

Democratic pragmatism in action

Democratic pragmatism is often proposed as a remedy for the crisis of administrative rationalism detailed in the previous chapter. On this account, there is nothing wrong with administration that a healthy dose of democracy cannot fix. This dose comes not in removing problem solving from administration and putting it in the hands of representative institutions such as legislatures; rather, it is a matter of making administration itself more democratic. This task can be accomplished in a variety of ways, many of which were pioneered in environmental policy (and spread to other policy areas).

In the previous chapter I noted that environmental administration is in crisis, indicated by diminishing returns to administrative effort. Democratic pragmatism can speak directly to this crisis, and later I will assess its performance in these terms. Some of the impetus for democratic pragmatism now comes from a desire to make administration more flexible and responsive to varied circumstances (Fiorino, 2004). But the main reason for the democratization of environmental administration has often been a felt need to secure legitimacy for decisions by involving a broader public. The most vivid case of a national legitimation crisis leading to outreach to participation by environmental interests occurs in the United States around 1970. The Nixon administration felt besieged by the anti-war movement, radical elements of the civil rights movement, and the counter-culture more generally. The administration reached out to environmentalists as the least radical elements of the counter-culture, and drew them into government (Dryzek et al., 2003: 57–60).

Several devices are available to democratic pragmatists, as follows.

Public consultation

One important item in the repertoire of administrative rationalism is environmental impact assessment, under which a statement is prepared detailing the anticipated impact of a project (be it a freeway, pipeline, or land-use plan) on the environment (see Chapter 4). Impact assessment is designed to force administrators to consider environmental values and scientific evidence that they might otherwise have excluded or overlooked. But impact assessment is invariably accompanied by opportunities for public comment on the document produced. Sometimes this is mere symbolism, if there is nothing to force the department in question to take into account the substance of public comment in its subsequent decision on the proposal. Still, policy makers must both anticipate and respond to comment. In the process set up in the United States under the National Environmental Policy Act of 1970, the responsible federal agency must produce a draft statement, release that document for comment, compile responses (from other government agencies, other levels of government, environmental and community groups, interested corporations, resource users, and ordinary citizens), and respond to these comments in the final version of the statement. Thus information from a variety of perspectives is systematically sought out. This information will rarely have direct, traceable impacts on agency decisions; but more subtly it may alter the context in which administrative decisions are made and implemented, by changing the discourse surrounding policy determination in a way that makes both environmental and democratic values more legitimate and more visible. The way Bartlett (1990) puts it, environmental impact assessment can constitute a 'worm in the brain' of the administrative state, one that moves it in simultaneously more democratic and more environmentally sensitive directions.

Public consultation can also proceed without being tied to particular documents such as impact statements. For example, several European countries (Sweden, the Netherlands, and Austria) initiated extensive consultative efforts in the late 1970s concerning the future of nuclear power (see Nelkin and Pollack, 1981). These exercises did not involve much in the way of transfer of power from the state to the citizenry. But they did have real consequences: for example, in 1979 the Swedish government decided not to construct any more reactors, and to begin phasing out nuclear energy.

Public consultation is now regarded as a necessary accompaniment to environmental policy development in many countries, and on no issue is this more true than genetically modified organisms in agriculture. In 2003 the UK government initiated 'GM Nation,' a six-week consultative exercise on genetically modified crops and foods. GM Nation involved more than 600 public meetings across the country, and publicization of scientific reviews and analyses of the costs and benefits of genetically modified organisms.[1] Around 36,000 questionnaires were turned in, and opinions could also be submitted via a website. The idea was to involve ordinary people, though self-selection of participants meant activists were over-represented. The summary reported public unease on both health and environmental grounds, together with widespread distrust of government and biotechnology companies. In 2004 Environment Minister Margaret Beckett announced the government's response. GM crops should be evaluated on a case-by-case basis under precautionary guidelines, and that GM food products should be clearly labeled.

Alternative dispute resolution

Opportunities for public comment do not formalize any particular role for nongovernmental participants. One way of recognizing and involving interested parties is through alternative dispute resolution (ADR). ADR has arisen in legalistic systems—notably the United States—as an alternative to the expensive stalemate that prolonged legal actions entail. The idea is to bring the parties to a dispute together under the auspices of a neutral third party (often a professional mediator) such that they might reason through their differences. The target is consensus under which all parties are better off than they would be in the absence of an agreement (Susskind et al., 2000). ADR reached the environmental realm in the 1970s under the heading of environmental mediation, and since then disputes have been mediated on a wide variety of issues, including complex ones. These issues include construction of dams, irrigation schemes, mines, shopping malls, and roads; watershed management; siting of hazardous waste disposal operations; clean-up of existing hazardous waste sites; ecological restoration; and anti-pollution measures. Mediation functions not just as an alternative to the courts. Government agencies can also use and sponsor it when encountering resistance to their proposals. The relevant participants might include community representatives, environmental groups, corporate developers or polluters, government departments, and local governments. Thus mediation can play a role in policy making rather than dispute resolution narrowly defined. Kai Lee (1993) believes that this is a productive way of channeling political conflict into administrative decisions. In particular, he believes ADR has an essential role in effective ecosystem management, providing creative ways for conflicts to end in learning rather than in victory for one side and defeat of the other. Other observers are more skeptical, seeing ADR mainly in terms of the co-option and neutralization of troublemakers (for example, Amy, 1987). It is perhaps best to accept that ADR has an ambivalent potential. At a minimum, it demonstrates that decisions must be legitimated through participatory procedures. These procedures can involve neutralization and co-option; but they can also involve democratic principles eating away at the administrative state, forcing it to open its ways. So it is up to democratic pragmatists and perhaps even proponents of more radical democracy to make the most of these cracks in the citadel of the administrative state and its legal accompaniments (Torgerson, 1990: 141–5).

Policy dialogue

Environmental mediation and other forms of ADR are often case-specific or site-specific. However, principles of reasoned discussion oriented to consensus can also be applied to more strategic policy issues, though the success rate (in terms of reaching an agreement and having it put into policy practice) is lower than for more circumscribed cases. An early example came with the National Coal Policy Project in the United States, which operated in the late 1970s to bring several national environmental groups and coal producers together to devise a strategy for coal mining and coal burning (McFarland, 1984). The environmentalists agreed to a simplified one-stop permitting process for new coal-burning power plants, and in return the coal producers agreed to public funding of environmental objectors to such plants. However, the recommendations of the project were never put into policy practice, in part because officials from government agencies with interests of their own were not included in the negotiations (nor were coal workers or their unions).

A clearer example of policy dialogue more explicitly connected to—indeed, sponsored and funded by—government may be found in Australia, with the Ecologically Sustainable Development (ESD) process initiated under Prime Minister Bob Hawke in 1990. ESD began with an invitation to the main national environmental groups and relevant industry representatives to participate in discussions oriented to the generation of strategic policy recommendations in a number of areas: agriculture, energy, fisheries, forests, manufacturing, mining, and tourism. In each area a working group was set up, and a report eventually produced. The four invited environmental groups were the Australian Conservation Foundation, World Wildlife Fund, Greenpeace, and the Wilderness Society. The Wilderness Society withdrew immediately due to its unhappiness with other government policies contrary to sustainability; later, Greenpeace withdrew. However, both groups remained in contact with the two groups that continued to participate. When the time came for the ESD groups to report, Hawke had been replaced by a prime minister committed to confrontation rather than consensus, who placed a much lower priority on environmental issues, which had also faded from public prominence with the arrival of economic recession. Thus few of the recommendations found their way into public policy.

A more effective translation of policy dialogue into policy practice may be found in the Canadian province of Alberta, the first place in North America to solve the NIMBY (Not In My Back Yard) problem for hazardous wastes. Nobody wants a hazardous waste treatment facility in their backyard. Given that everywhere is someone's backyard, and given the relative ease of access to veto power in the Canadian and US political systems (reinforced in the United States by the prominent role of the courts), the normal condition of policy on this issue is impasse. Recognizing this problem, the government of Alberta in the late 1980s initiated a dialogue with local community groups and industry, which eventually produced consensus on a site and principles for its development and operation. The process involved a referendum on the basic idea of siting, funding to communities to employ experts, regular seminars and public meetings. Once the Swan Hills site was selected and the treatment plant built, communities in the vicinity received further funding and access to monitoring reports and expert advice (for details, see Fischer, 1993: 176–7; Rabe, 1991). Unfortunately the operation of Swan Hills proved problematic, re-creating a familiar standoff of environmentalists and indigenous peoples against operators of the site.

Localized policy dialogues have occurred all over the world under the 'Local Agenda 21' initiative coming out of the United Nations Conference on Environment and Development in 1992, which encouraged local governments to develop sustainability plans. These dialogues vary enormously in their process and substantive coverage, though most provide opportunities for citizen participation (Mason, 1999).

Lay citizen deliberation

ADR and other policy dialogues normally involve partisans, be they environmentalists or developers. But reflective deliberation can actually be achieved more easily by non-partisan lay citizens, who are more open to argument and persuasion. This capacity helps explain the rise of deliberative exercises built around lay citizen participation (Carson and Martin, 1999). Examples include consensus conferences (invented in Denmark), citizens' juries (which began in the United States but have been used most extensively in the United Kingdom), deliberative opinion polls (created by James Fishkin), and planning cells (found in Germany). The number of participants varies from around fifteen in a citizens' jury to several hundred in a deliberative poll or planning cell. Participants are recruited through random selection or some stratified sampling procedure, and brought together for two days or more to deliberate the issue in question. There are opportunities to question experts and advocates, and plenty of information is provided. The product is normally a policy recommendation (though in deliberative polls, participants simply fill out a questionnaire at the end). Lay citizen deliberations have occurred on issues encompassing genetically modified organisms in agriculture (in many countries), container deposit legislation in Australia, energy policy in Texas, urban planning in Germany, and wetlands protection in England.

Public inquiries

Public inquiries resemble impact assessment in that they are oriented by a specific project proposal. But rather than just producing a document and allowing public inspection and comment, a public inquiry involves a visible forum where proponents and objectors alike can make depositions and arguments. Much depends upon the terms of reference, and the way these terms are interpreted by the individual presiding over the inquiry. The terms and their interpretation can be narrow and biased toward the project proponent. This is how inquiries into proposed nuclear installations in Britain typically proceed. So Kemp (1985) chronicles the case of a 1977 inquiry into a Thermal Oxide Reprocessing Plant (THORP) proposed for Windscale in Northwest England (then, as now, a notorious site of radioactive pollution). The project proponent, British Nuclear Fuels Ltd, was allowed to introduce evidence on the economic benefits of THORP, but objectors were not allowed to bring economic evidence against it. The legalistic rules of the inquiry were congenial to the well-funded proponents, not to the resource-poor objectors; and the proponents could deploy the Official Secrets Act at key points. Not surprisingly, Mr Justice Parker presiding over the inquiry came down in favor of THORP. Contrast this with the contemporaneous inquiry into proposed oil and gas pipeline construction from the Arctic to Southern markets conducted in Canada by Mr Justice Thomas Berger. Berger took pains to make sure that resource-poor interests, especially indigenous peoples, were provided with funds, access to expertise, and an ability to testify in a forum under conditions with which they were familiar (the inquiry traveled to remote villages). He interpreted the terms of reference broadly, to encompass development strategies for the Canadian North, not just whether or not pipelines should be built. In this sense, the inquiry became more like a policy dialogue. Berger's report (Berger, 1977) proposes a reinvigorated renewable-resource based economy for the Canadian North, in which oil and gas development have little place. Berger pushed democratic pragmatism to its limits—and perhaps beyond, to the kind of participatory process favored by green radicalism.

Another Canadian case that resembles Berger but was styled a 'scientific panel' rather than 'public inquiry' can be found in the mid-1990s in Clayoquot Sound in British Columbia, where the issue involved logging. The panel was set up to resolve an impasse between logging companies on one side and environmentalists and indigenous people on the other. Science was interpreted broadly to encompass and integrate the traditional ecological knowledge of the Nuu-Chah-Nulth people, and a Nuu-Chah-Nulth chief chaired the panel. The panel shifted policy concern from economic efficiency to ecological integrity (Torgerson, 2003: 128–31). While the Berger and Clayoquot Sound cases provide encouragement to ecological democrats, their rarity (even in Canada) should give pause for thought.

Right-to-know legislation

If individuals from outside government are to be effective participants in democratic processes they need access to relevant information. Sometimes access will be facilitated by general freedom of information laws under which governments operate. These apply to some governments more than others. For example, the British counterpart to freedom of information has long been the Official Secrets Act, which presumes that everything is secret if it has the remotest connection to national security (this comes into play, for example, on anything relating to nuclear power). More specific to environmental politics is right-to-know legislation which specifies that corporations must disclose information relating to (say) the risks to workers of particular chemicals, the routes and timetables of shipments of noxious substances, and the toxicity of wastes being stored, transported, and dumped. Such laws exist in a number of Canadian provinces and US states.

These six developments all involve injections of democratic pragmatism into the administrative state. In every case experience has been mixed, and a lot of skepticism remains, especially from those committed to more radical expression of both environmental and democratic values. But in some cases, notably Berger and Clayoquot, we can glimpse the possibility of a more radically participatory and discursive democracy. At any rate, taken together, these six developments indicate the degree to which environmental policy has over the past forty years been accompanied by greater openness and participation in decision making. As Paehlke (1988) notes, all this is a far cry from the gloomy prognostications of survivalists who argued that environmental limits could only be confronted by centralized and authoritarian government. Of course, survivalists might still say that all the policy effort of the last four decades has not really confronted the issue of environmental limits head-on, and that we are still on course for overshoot and collapse.

Democratic pragmatism as government and governance

Democratic pragmatism describes an orientation to governing in its entirety, not just the inspiration for a variety of specific reforms and exercises of the kind described in the preceding section. This orientation stresses interactive problem solving involving participants from within government and outside it. Such interaction can occur in the context of committee meetings, legislative debate, hearings, public addresses, legal disputes, rule-making, project development, media investigations, and policy implementation and enforcement; it can involve lobbying, arguing, advising, strategizing, bargaining, informing, publishing, exposing, deceiving, image-building, insulting, and questioning. In this light, the real stuff of government is not to be found in constitutions and formal divisions of responsibility. Rather, it is to be found in interactions that are only loosely constrained by formal rules. Quiet conversations in the bar may matter as much as speeches to parliament.

These interactions occur whether or not constitutions, laws, rules, and organization charts say they should. Thus democratic pragmatism can readily assimilate the kind of decentralized networked governance discussed as a corrective to administrative rationalism in Chapter 4.

'Government' has a top-down imagery in which administrative rationalism rules once goals and principles have been set. 'Governance,' in contrast, dispenses with a central locus of authority, reveling in informal interactions. The actors involved might be public officials, but they might equally well be nongovernmental organizations (NGOs), lobbyists, activists, journalists, corporations, international organizations, or governments in different jurisdictions. The image of the network replaces that of the hierarchy (Castells, 1996). A network has multiple nodes and complex pathways of interactions across participants. Communication in a hierarchy goes up to and down from the apex. In a network, public outcomes can be produced without the stamp of any apex, or even without the approval of any government. This happens when a consumer boycott forces a corporation to change its ways (for example, to catch tuna in a way that does not endanger marine mammals and turtles); or when activists and businesses agree on a corporate code of conduct. Jänicke (1996) calls such actions 'paragovernmental.'

In the 1990s, the Reinventing Government task force (charged with looking at the structure of policy delivery) headed by US Vice-President Al Gore proposed reforms to the EPA's regulatory system along 'governance' lines, to emphasize public–private partnerships and collaborative relationships across regulators, communities, and businesses. Republicans in Congress eager to devolve power from the federal level were supportive. Environmentalists were less eager, remembering perhaps that in federal systems it is normally higher levels of government that are more committed to environmental values. Still, the United States in the 1990s saw substantial impetus toward more flexible, decentralized, and cooperative environmental regulation (Fiorino, 2004).

In both the United States and the European Union, networked governance became associated with 'new environmental policy instruments' (Jordan et al., 2003). Some of these instruments are market-based (and so will be discussed in Chapter 6). Others are more authentically collaborative governance measures. These include voluntary agreements between regulators and corporations, and environmental management and audit systems under which corporations voluntarily set targets for their own environmental impacts and monitor progress, receiving a stamp of approval if targets are met. This stamp does not have to be given by government. Transnational forest certification for wood products is administered by a network of NGOs (including the Rainforest Alliance and World Wide Fund for Nature) and corporations, coordinated by the Forest Stewardship Council, founded in 1993 (Meidinger, 2003). Collaborative governance can also be found in natural resource management. Innes and Booher (2003) describe the case of the Sacramento Water Forum, which seeks consensus encompassing businesses, environmentalists, farmers, and local governments in an ecologically degraded setting.

Skeptics would see in such ventures an abdication of public authority into private hands, and plenty of opportunity for businesses to engage in symbolic 'greenwashing' of their activities. Those distrusting business power would stress the need for governance to be exercised over corporations, not just with them. Along these lines, Braithwaite and Drahos (2000) show how activists and NGOs as well as states have helped to create a global governance web that regulates business beyond the level of the nation-state and beyond the reach of conventional administrative rationalism.

Political interaction can, then, involve complex paths of communication. To the administrative rationalist, this might sound like chaos and subversion. But, arguably, this apparent chaos has its own rationality, what Charles Lindblom has called 'the science of muddling through' (1959), or 'the intelligence of democracy' (1965). This 'science' is the opposite of administrative science, for it revels in unclear divisions of responsibility, political conflict, bending the formal rules so as to make things work, and substituting ordinary knowledge for analysis. Problems are solved piecemeal, usually through series of compromises among the different actors concerned with an issue. Interaction substitutes for analysis; different actors bring different perspectives and concerns, which are somehow agglomerated into policy decisions.[2]

Pragmatists believe such processes are the best means for attacking public problems. A justification for the essential rationality of decentralized political interaction (though only within the confines of the liberal democratic state) has also been advanced quite famously by Karl Popper (1966).[3] Popper's model problem-solving community is found in successful sciences, where the rational attitude is to advance theories capable of being put to the test, then seek criticism of them through as many tests as possible, especially experiments. Popper believes that this attitude should apply in politics and policy making too. Public policies are like experiments. Nobody knows in advance if a particular policy will succeed or fail. So it should be tried first on a limited scale, and reactions sought from as many different directions as possible about its positive and negative effects. Popper calls this kind of policy making 'piecemeal social engineering.' The only way to ensure feedback from as many different directions as possible is to have policy proceed in a liberal democratic setting, where different interests and actors (such as environmental and community groups, professional associations, different kinds of scientists, elected representatives, corporations and their officials, labor unions, and journalists) are all able to give their opinions without fear. Real-world liberal democracies and governance networks are only imperfect approximations to Popper's 'open society' ideal, but their defenders would say no closer style of politics has yet been found.

Is this sort of policy making by interaction appropriate to an ecological context? The apparent chaos of piecemeal, interactive politics might belie a deeper organization, even if nobody in the system perceives it or really understands why and how this organization happens. Interactive politics in one sense resembles ecosystems, for both are self-organizing systems (diZerega, 1993). That is, complex structures of order evolve without anyone designing them, as a result of the relatively simple and shortsighted choices and actions of individual organisms. In this light, the real order of democratic pragmatism is not to be found in constitutions, but in informal, interactive processes. Of course, the precise structure of order in any self-organizing system matters a great deal. By definition, an environmentalist can have little quarrel with the kind of order that ecosystems have produced by evolution. That liberal democracy and governance networks are self-organizing systems (as is the capitalist market) does not mean they are adequate in light of ecological criteria.

Ecosystems are self-organizing systems full of negative feedback devices that correct for disturbances. For example, a forest fire is normally followed by pioneer species of plants springing up in the burnt area, which in turn provide the growing conditions for more mature forest species to return. The idea of negative feedback also defines the metaphor of the thermostat. So what kind of 'thermostat' does democratic pragmatism possess? The answer lies in the variety of individuals, organizations, parties, and movements which can bring pressure to bear in and on political interaction in response to environmental disturbances. For example, one would expect wilderness advocates to keep their eyes on old growth forests, so that if clear-cutting threatens to get out of hand they can protest, lobby, hold press conferences, issue legal challenges, and so forth. Or if a proposal for a toxic waste incinerator in an urban neighborhood threatens life and health, the local community can organize against it. Survivalists would argue that these kinds of actions are reactive and so incapable of anticipating limits before we hit them, though survivalist groups such as Zero Population Growth also participate in liberal democratic politics.

Whether or not these negative feedback devices are ecologically adequate depends crucially on the values of the people through whom the devices act. If people value tangible material goods above all else, then feedback will be impaired (though even these individuals might protest against any immediate environmental threat to their life and health). So is there anything in democratic pragmatism intrinsically conducive to ecological values? Two arguments claim that there is.

Democratic pragmatism involves talk and written communication, not just strategizing and power-plays, and such communication works best when it is couched in the language of the public interest, rather than private interests. Steven Kelman (1987) believes such talk is not cheap, and that people actually internalize public interest motivations. Adolf Gundersen (1995) applies this sort of analysis to public deliberation about environmental affairs. Deliberation is necessary for democratic pragmatism to work. Problem solving in democratic pragmatism, recall, is never a matter of individuals acting in isolation or under command from anyone else. Instead, problems always get discussed. Gundersen believes the very act of discussion or deliberation about issues activates commitment to environmental values, or, more precisely, 'collective, holistic, and long term thinking.' Gundersen's evidence is a series of forty-six 'deliberative interviews' he conducted with a variety of people who did not in the beginning identify as environmentalists. By the end of these discussions, all espoused environmental values more strongly. On this account, everyone has latent positive dispositions which only need to be activated into specific policy commitments. Discussion in democratic settings forces people to scrutinize their own dispositions in a way that promotes such activation.

The idea that participation in democratic settings activates environmental values is shared by Mark Sagoff (1988). Sagoff believes that every individual has two kinds of preferences: as a consumer and as a citizen. These preferences may point in quite different directions for the same individual. His running example concerns the Mineral King Valley in California's Sierra Nevadas, where the Walt Disney Corporation wanted to build a ski resort. Confronting his students with this possibility, it turns out that many of them would enjoy visiting such a resort to ski and enjoy the après-ski nightlife. Few had any interest in backpacking into the existing Mineral King wilderness. But when asked whether they would favor construction of the resort, none did. The answer is that while as consumers they would love to ski there, as citizens they object to wilderness destruction. Citizen preferences are more concerned with collective, community-oriented values, as opposed to the selfish materialism of consumer values. While one might dispute the degree to which such public-spirited motivation pervades real-world liberal democratic politics, Sagoff's critique of economic reasoning and market rationality as applied to environmental policy is devastating. He also deploys his argument to excuse some of his more disgusting personal habits, notably driving a car that leaks oil everywhere which sports an 'ecology now' bumper sticker (Sagoff, 1988: 53). The sticker proclaims his citizen preferences, the oil slick under his car his consumer preferences. The citizen in him would like the government to crack down on the consumer in him.

Discourse analysis of democratic pragmatism

Basic entities whose existence is recognized or constructed

Like administrative rationalism, democratic pragmatism takes the structural status quo of liberal capitalism as given. However, the treatment of government is very different. Government and governance are not treated as monoliths, but rather in terms of a multiplicity of decision processes populated in large part by citizens. *Homo civicus* figures large, *homo bureaucratis* hardly at all. Democratic pragmatism has little or nothing to say about ecosystems and the natural world; very different conceptions on this score are welcome in debate.

Assumptions about natural relationships

Both administrative rationalism and democratic pragmatism place nature as subordinate to human problem-solving efforts. Whether nature contains self-regulating ecosystems or is just a storehouse of brute matter and energy makes little difference here. The natural relationships within human society postulated by the two discourses are in contrast quite different. Democratic pragmatism celebrates equality among citizens (of course, the reality of liberal democracy may be very different). Everyone has the right to exert political pressure, be they scientists, elected officials, pressure group leaders, ordinary voters, or ordinary non-voters. Beyond this basic equality, political relationships are seen as interactive and far more complex than those in a bureaucratic hierarchy. Interactions feature a mix of competition and cooperation. Certainly cooperative problem solving can occur; but so can political conflict between partisans of competing interests (such as environmentalists and developers).

Agents and their motives

Agency in democratic pragmatism is for everyone, be they individual citizens and political activists or collective actors such as corporations, labor unions, environmentalist groups, community organizations, and government agencies. Motives are mixed. Starting from theory rather than practice, Wissenburg (1998) argues that the building blocks of green liberalism may be found in agents who accept 'ecoduties' as part of their basic social responsibilities. Individuals must also accept a 'restraint principle' whereby any environmental damage they do cause requires restoration or compensation on their part. But this specification of principles begs the question of how they might come to the fore. (The obligations that might accompany ecological citizenship are further developed by Dobson (2004), who stresses sustainability in the ecological footprint one imposes on the Earth.)

Many actors much of the time pursue selfish material interests, such as profit, increased property values, higher wages, more secure employment, or subsidized access to a favorite natural area. But the discourse of democratic pragmatism requires that at key junctures agents can be motivated by the public interest, and recognize that there are community interests (such as ecological integrity) that transcend individual interests (de-Shalit, 2000: 92–129). In the first instance, the public interest will have to be defined in plural terms. So what the Wilderness Society takes as the public interest will not necessarily be the same as the Chamber of Commerce. Some democratic pragmatists, especially those attuned to networked governance, would leave it at that, arguing that plurality here is irreducible, and that we can expect only piecemeal compromises across partisans of different views. But others, such as Kelman, Gundersen, and Sagoff, hope reasoned public dialogue will produce convergence on a common conception of the public interest (see also Williams and Matheny, 1995). If a single public interest does emerge through dialogue, it is a very different matter from the unitary public interest that exists for administrative rationalists, for whom the public interest is something for analysts to discover, rather than the public to debate. So 'environmental sustainability' can be treated as a scientific concept to be administered; or as something to be explored in democratic debate (Arias-Maldonado, 2000).

Key metaphors and other rhetorical devices

Two scientific metaphors are advanced by reflective defenders of democratic pragmatism. The first draws from physics to treat public policy as the result of forces acting upon it from different directions. These forces differ in the direction in which they want to pull public policy, and in their relative power. Such a metaphor is likely to be employed by those who think there can be no unitary public interest. This metaphor was long a staple of pluralist accounts of the US political system developed by American political scientists.

A second metaphor is that of science in its entirety. As we have seen, Popperians believe that public policies are like scientific experiments, and that the proper attitude for scientists and policy makers alike is an open, critical, and democratic one.

Another metaphor I mentioned earlier is that of the thermostat, designed to trigger interventions (heating and cooling) as soon as temperature departs from a desirable range. Democratic pragmatism allows attention to a wide range of target variables analogous to temperature (economic and political as well as environmental ones), and many ways in which negative feedback can be brought to bear. Foremost among these is the possibility for aggrieved citizens and groups to mobilize when they perceive an environmental abuse.

Finally, the network itself is a kind of metaphor, especially when deployed by those who emphasize the information society in which we live. Parallels are drawn between networked information technology (especially the internet) and networked governance. Both proceed without any central controller.

The limits of democratic pragmatism

Democratic pragmatism has much to be said on its behalf. It accepts many problems that baffle administrative rationalism. This transfer is often made for reasons relating to the need to legitimate policy decisions in the eyes of a broader public, but it can be justified in terms of more effectively resolving problems too. If we look around today's world, we see that the countries that have progressed most in terms of environmental conservation and pollution control are the ones where democratic pragmatism is most common: the capitalist democracies (though the most capitalist are not the best performers). The latter piece of evidence does not provide

BOX 5.1 Discourse analysis of democratic pragmatism

1. **Basic entities recognized or constructed**
   - Liberal capitalism
   - Citizens
2. **Assumptions about natural relationships**
   - Equality among citizens
   - Interactive political relationships, mixing competition and cooperation
3. **Agents and their motives**
   - Many different agents
   - Motivation a mix of material self-interest and multiple conceptions of public interest
4. **Key metaphors and other rhetorical devices**
   - Public policy as a resultant of forces
   - Policy like scientific experimentation
   - Thermostat
   - Network

quite the comfort to democratic pragmatism than it might. For the acknowledged leaders in the environmental stakes include countries such as Germany and Japan (Scruggs, 2001), where there are limits on who can have access to policy making and under what terms. So in Japan policy is monopolized by business and government elites; in Germany, labor union leaders also have a say. In each of these cases, usually described as corporatism, participation is through highly formalized channels, allowing little of the self-organizing give and take celebrated by democratic pragmatists. Moreover, some of the best-performing countries are adopting a discourse quite different from democratic pragmatism, as we will see in Chapter 8.

Prometheans argue that it is the prosperity of the developed capitalist democracies that allows them to cope better with their environmental problems than anyone else, as opposed to the intrinsic problem-solving qualities of democratic pragmatism. And there is always the possibility that these countries have offloaded many of their environmental problems onto poorer countries. So a clean and pleasant environment in Japan is purchased in part by the dirtier elements of manufacturing industry being transferred to other East Asian countries, not to mention deforestation of Southeast Asia to meet Japanese timber needs.

The main limit to democratic pragmatism is the simple existence of political power (which goes unrecognized by enthusiasts such as Gundersen and Sagoff). Politics in capitalist democratic settings is rarely about disinterested and public-spirited problem solving in which many perspectives are brought to bear with equal weight. Often there are powerful interests with large financial resources at their disposal that try to skew the outcomes of policy debates and decision-making processes in their direction (González, 2001*a*). Sometimes that direction will coincide with ecological values. More often it will not, as the interests with the greatest amount of resources and the strongest incentives to deploy them are business interests.

Business can influence the terms of debate by producing glossy advertising material to tout the environmental friendliness of its products. It can sponsor Earth Day festivities. It can produce television advertising to promote the corporate environmental image: so in the United States Weyerhauser once promoted itself as 'the tree growing company,' with film of a bald eagle flying over a forest. The clear-cutting of old growth forests, which is also one of Weyerhauser's activities, is unmentioned. Corporate actors also have greater access to expert counsel in public inquiries. Alternative dispute resolution can be manipulated by these actors and their sympathizers in government in order to co-opt and neutralize troublemakers from community and environmental groups. ADR can be oriented toward a 'responsible development' gloss on projects which generally go ahead, and toward treatment of environmental values as on a par with business's material interests (Amy, 1987). Participation by environmentalists in impact assessment might dissipate energies that would be better spent on other activities, if the process merely legitimates decisions already made elsewhere on the basis of economic values or corporate profit (Amy, 1990: 60–4). Corporations can even offer employment to environmental activists. For example, leading British Green Jonathan Porritt signed on as an advisor to Sainsbury's, the food retailing giant.

Such pressures do not go all one way; public opinion does exist as something more than the creation of business public relations departments, and public interest groups can mobilize expertise and support, even money. Still, so long as the structural status quo of the capitalist market economy is taken as given, business has a 'privileged' position in policy making, for government relies greatly upon business to carry out basic functions such as employing people and organizing the economy (Lindblom, 1977: 171–5). Any measures for environmental protection, conservation, or pollution control which threaten to undermine business confidence will be automatically punished by disinvestment. This possibility casts a long shadow over policy deliberations, however democratic they may be (see Press, 1994). And once business publicists realize this, they can make good strategic use of the disinvestment threat, even when there is no real intention to disinvest. As we will see in Part IV, sustainability discourse dissolves such problems by eliminating the conflict between economic and environmental values.

Democratic pragmatism as a discourse recognizes citizens as a basic entity, and a natural relationship of equality across citizens. But this imagery of reasoned debate among equals is in practice highly distorted by the exercise of power and strategy, and by the overarching need of government to maintain business confidence. Matters appear still more doubtful in an ecological light when one further considers the character of actors and their interests. One advantage of democratic pragmatism stressed by its adherents is that it enables views on policy proposals to come from a variety of directions. Some directions represent conceptions of what is in the public interest. These conceptions may vary: to some, the public interest may involve mostly economic efficiency, to others distributional equity in society, to others still ecological integrity, to others social harmony. When enthusiasts such as Gundersen and Sagoff in an ecological context, and Dewey and Popper more generally, think of democratic debate, this is presumably what they have in mind.[4] But other interests involved are motivated mostly by their own material interests: corporations and industry associations concerned with maximizing profit and avoiding environmental controls on their operations, or labor unions concerned with the income and employment of their members, even if that means employment in unsustainable practices such as clear-cutting of ancient forests. The pluralist aspect of democratic pragmatism treats all such interests and concerns as equally legitimate (see Williams and Matheny, 1995: 19–24). The mere fact of participation in liberal democratic settings does not lead actors to discard their motivations as consumers and producers in favor of more public-spirited citizen preferences, or to conclude that pursuit of their economic interests should be confined to the market place rather than allowed to enter politics.

More insidious still are special interests that masquerade as general principles. So, for example, the 'Wise Use' movement in the American West in the 1990s has a name that connotes commitment to sensible use of resources, but in practice it seeks a regime of subsidized access for local communities and corporations to minerals, grazing rights, and timber located on public lands in the region.

Political rationality in democratic pragmatism means that all actors have to be mollified, pretty much in proportion to their ability to create difficulties for government officials, irrespective of whether they are motivated by conceptions of the public interest or more selfish material interests. This does not necessarily coincide with ecological rationality, which is concerned with the integrity of natural life-support systems (see Dryzek, 1987: 118–20). So in 1993 the Clinton administration took a small step toward ecological rationality when Secretary of the Interior Bruce Babbit proposed reforming grazing law to end subsidized access for cattle ranchers to public land. It soon became evident that the politically rational thing to do was back off on these reforms for fear of the electoral weight of the western states where these reforms would take effect, and where welfare ranchers and their sympathizers could tip the balance come election day.

Democratic pragmatism in some respects merits a similar summary judgment to administrative rationalism: plenty of achievements to look back on, but limits to effectiveness increasingly apparent. This similarity applies mostly at the level of specific policies and institutions inspired or justified by the two discourses. But as a discourse, democratic pragmatism has one striking advantage: it is more conducive to an awareness of the limitations of its own institutional manifestations, and so to efforts to overcome these limits.

NOTES

1 See **www.gmnation.org.uk.**

2 There are numerous case studies of this kind of process in the political science literature. Classics may be found in the work of Aaron Wildavsky (Wildavsky, 1988; Pressman and Wildavsky, 1973).

3 Popper differs somewhat from the pragmatists in believing that there can be general laws of nature and of society which natural scientists and social scientists alike can discover. Pragmatists believe only that there are particular problems to be solved, not laws to be discovered.

4 For a more explicit statement about the degree to which 'public spirit' actually pervades even US politics, see Kelman (1987). According to Kelman, presidents, congresspersons, and bureaucrats alike are all motivated mainly by the desire to make an honest effort to achieve good public policy. But even Kelman recognizes that special interests will sometimes upset this happy situation.

6

Leave it to the Market: Economic Rationalism

When it comes to theories to guide public policies, democratic pragmatists are quite agnostic. The only test they are inclined to apply is the pragmatic one of whether the policy inspired by the theory works out in reality. Politics of the sort pragmatists favor is home to believers in many different theories and perspectives. In the last three decades, the most prominent perspective on policy in general has been an economic one. This perspective goes by different names in different places: market liberalism, classical liberalism, neoliberalism, and free-market conservatism. Sometimes it is even personalized, and becomes Thatcherism in the UK, Reagonomics in the United States, Salinastroika in Mexico (after President Salinas de Gotari), or Rogernomics in New Zealand (for finance minister Roger Douglas). Now, many of those who sail under these banners are Prometheans, who believe that the only task for government in environmental affairs is to leave markets well alone, such that human ingenuity can be given full rein. Yet there are others just as committed to market principles, who recognize that, whatever the case in other areas, markets in environmental goods do not always exist, and so often need to be created and managed, sometimes even by taxation. Thus their discourse is rationalistic, entailing substantial cogitation, calculation, and design on the part of policy makers.

Economic rationalism may be defined by its commitment to the intelligent deployment of market mechanisms to achieve public ends. It differs from administrative rationalism in its hostility to environmental management on the part of government administrators—except, of course, in establishing the basic parameters of designed markets. In this one key aspect economic rationalism turns out to depend on the administrative rationalism it otherwise so despises. The commitment to markets might imply that economic rationalism's natural political home is on the political right. In the United States, 'free market environmentalists' revel in this connection, almost certainly to the detriment of the policies they favor. Yet some people with left and/or green credentials are also attracted by the use of markets in an environmental context (for example, Daly, 1992; Roodman, 1996), and countries with governments organized along more social democratic lines (for example, Germany, the Netherlands, and France) have pioneered economic rationalist environmental policy instruments. One particular instrument (the congestion charge) is associated with socialist London Mayor Ken Livingstone.

William Reilly, Administrator of the EPA under President George Bush the elder and before that head of the Conservation Foundation, declared that 'The forces of the marketplace are powerful tools for changing individual and institutional behavior. If set up correctly, they can achieve or surpass environmental objectives at less costs and with less opposition than traditional regulatory approaches' (quoted in Yandle, 1993: 188). Changes of president did not dampen this enthusiasm: in 1992 president-elect Bill Clinton spoke of 'harnessing market forces' to induce companies to incorporate 'environmental incentives into daily production decisions' (quoted in Nelson, 1993: 1). In 2003 George W. Bush's EPA Administrator Mike Leavitt proclaimed his belief in 'markets before mandates,' as 'market-based approaches and economic incentives often result in more efficiency at less cost.'[1] In 2004 US Interior Secretary Gale Norton described a 'new environmentalism' under which landowners would get incentives to conserve habitat, though this turned out to involve grants and subsidies of the kind economic rationalists dislike.

Market-type policy instruments have been promoted by the Organization for Economic Cooperation and Development, the rich man's club of the world's developed countries (see OECD, 1989), and the European Union. The European Environment Agency (2000: 397) has even proposed a comprehensive regime of environmental taxes to replace income taxation as a main source of government revenue. Economic instruments were endorsed by the 1987 Brundtland Report, *Our Common Future*, which launched the era of sustainable development on the international stage. With all this enthusiasm it is paradoxical that the use of market-based instruments remains limited.

My discussion of economic rationalism begins with its purest strain, emphasizing the conversion of environmental resources to private property. I then move to less radical strands that stress market incentives but not necessarily private property.

Privatizing everything if you can

Markets are systems in which goods, services, and financial instruments are exchanged for each other. Markets work smoothly to the extent that participants in transactions can be confident that they do in fact have a right to sell or buy the goods in question—in other words, they have property rights, be it to a car, a can of beans, a company, a bond, or a piece of land. If we are to have markets in environmental goods, then we need private property rights here too. According to economic rationalism, specification and enforcement of these rights is the main task of government. Why are private property rights and markets so desirable? Because people tend to care more for what they hold privately than for what they hold in common with others. This is why (for example) there is more litter in public parks than in private yards, or why public grazing land in the American West is more degraded than private land. The metaphor of the commons plays a much smaller role than it does in survivalism, but economic rationalism has a clear solution to the tragedy of the commons: divide it into chunks of private property. Once the commons is divided, these chunks can be bought and sold according to who is prepared to pay the most. Economic rationalists tell us that, given a few assumptions, markets maximize social welfare; and markets in environmental goods should be no exception. The private property right to the good in question will be bought by whoever values it most, and can make the most profitable use of it.

It is easy to see how this logic of property and markets works for ordinary material goods, services, education, even human labor power; it is much less easy to see how it applies to the environment. But economic rationalists see no real difficulties in applying the same logic here. Meiners and Yandle (1993: viii) believe that 'environmental controversies seem to boil down to arguments about property rights.' Of itself, this interpretation does not imply that it is going to be easy to specify, enforce, and adjudicate an appropriate set of rights. Yet economic rationalists are adamant that failure to do so lies at the heart of environmental problems; as Mitchell and Simmons (1994: 148) put it, 'environmental problems must be understood more as failures by government to specify property rights than as offshoots of private profit-seeking.'

What, then does an appropriate set of property rights look like? When it comes to land, the answer is easy, as systems of private property are well established, and only need extending to all land. This is really only a political issue in countries with large amounts of land in public ownership, such as the United States. American free-market environmentalists are obsessed with the public lands issue. Much of this land is in the western states, and most of it is controlled by agencies of the federal government, especially the National Park Service, the Defense Department, the Forest Service, and the Bureau of Land Management. With the exception of the Pentagon, these are notionally professional land management agencies. In practice, as economic rationalists argue, these agencies often act as conduits for the abuse of land at the hands of special interests (see Anderson and Leal, 1991: 51–9, for a catalogue). Ranchers can graze their cattle on public land at below-market prices, and have little incentive to care for this land, because they do not own it. Logging companies gain heavily subsidized access to national forests, as the Forest Service constructs roads at public expense. Often, the Forest Service receives less money for a timber lease than it pays to construct roads into the lease area. This amounts to publicly subsidized wilderness destruction. Wilderness lovers for their part get free and often subsidized access to the back country, leading to overuse and degradation. Tourists get heavily subsidized roads and facilities in the more accessible parts of national parks, which again become overused and abused. Mining companies can make use of antiquated nineteenth-century laws that allow them to stake claims to minerals on public lands while paying virtually nothing.

According to economic rationalists, none of these abuses would occur if the land were privately owned. Ranchers would have every incentive not to overgraze, and to invest in soil and vegetation conservation. Owners of forests that could not be logged economically would keep them as wilderness areas, or invest in wildlife conservation in order to attract hunters or photographers, who would be charged admission to provide income for these conservation investments. If mineral rights were privatized, there would be a more orderly and efficient market in mines, rather than an inefficient scramble to make (subsidized) claims. If parks were privatized, tourists and backpackers alike would have to pay the market price for access, and private owners would again have every incentive to use the income to enhance the recreational value of the park. If anyone wanted to preserve wilderness for its own sake or for the sake of the species that inhabit it rather than for recreational opportunities, then they could buy it and do so. This is exactly what the Nature Conservancy, a private organization, currently does. The Nature Conservancy's wholehearted acceptance of market logic and an associated corporate model eventually led it to allow oil drilling and logging on some of its lands, to sell land to supporters for the construction of large houses, and to sell land to developers for luxury vacation houses (on Martha's Vineyard island, Massachusetts). These actions could be justified on the grounds they raised money to buy more land to protect, though they hardly looked like nature conservation.

Privatization of land is a major issue only in North America, because in most other developed countries most land is already private. Not so air and water, and here a little more ingenuity in the specification of private property rights is called for. The argument becomes applicable to more countries, because all of them have polluted air and water.

Air is hard to privatize because, of course, it moves around in the way land does not (barring the occasional earthquake, landslide, or soil erosion). But the useful properties of air can have private property rights attached to them, normally in conjunction with a parcel of land. So a right to breathe clean air can be attached to ownership or occupation of a piece of residential or commercial land. Anyone violating that right by emitting pollutants into the atmosphere can be pursued through the courts to either secure compensation or prevent violation of the property right to clean air. The legal system would play an expanded role in any such regime.

The immediate problem here is one of identifying polluters and tracing the effects of pollution on human health. This can be extraordinarily difficult, especially when there are multiple polluters. The air in my garden may not be clean, but am I coughing because of the methane given off by the nearby landfill, the heavy metals emitted from a local toxic waste incinerator, smog coming from car exhausts, or sulfur dioxide from the city's coal-burning power station? Or is it because my neighbor is burning her garbage? Clearly what is needed here are vast improvements in monitoring technology, and until that technology arrives, it is not surprising that property rights in air have made little headway. Market zealots such as Anderson and Leal (1991: 165–6) recognize this problem, which is why they fantasize about adding tracers to all pollution sources, and about advances in lasimetrics and satellite tracking of atmospheric chemicals.

Some of the same problems apply when it comes to water, though private property rights to clean water have been established in some cases. In Britain the private recreational fishing rights attached to a stretch of river or lakeside come with a right to water clean enough for fish to flourish. So any polluter, upstream or elsewhere in the lake, can be sued by the individual or fishing club holding the fishing rights. The Anglers' Cooperative Association has been zealous in bringing cases against polluters. The result is that British waterways, while rarely pristine, are much cleaner than they would be otherwise. Those benefiting include not just the fish and the anglers, but also the (very few, given the climate) people who swim in rivers and lakes, the (more numerous) people who rely on rivers as sources of drinking water, and the plants and animals of aquatic communities.

In arid regions, the main water issue concerns not pollution, but supply. Again the Western US offers the most contentious and troubled cases. Water rights to stream flows generally went to the first person to claim them. Thereafter, the 'use it or lose it' doctrine applies, which means that users must waste water when they do not need it, for fear of losing their right to it once they do need it. It does not require an economic genius to realize it would be more efficient to allow individuals and corporations to buy and sell rights to particular portions of the flow of a river or creek. Water policy in the US is also blasted by free marketeers for its enormous degree of public subsidy for questionable schemes that build dams and canals to supply agribusiness corporations and a few other wealthy interests. The main villain here is the Bureau of Reclamation, long regarded as one of the more powerful empires in the US government (see Reisner, 1993). The Bureau's aim is to create agriculture in the desert. The massive ecological costs of its efforts include elimination of stream flows, siltation behind dams, and soil salination. None of these public subsidies and ecological costs would apply if western agribusiness, cities, and industries had to pay market prices for the water they consumed. Welfare irrigation in this light is no more defensible than welfare logging, welfare ranching, welfare backpacking, welfare tourism, and welfare mining. All are costly and environmentally devastating.

Land, air, and water together cover a lot of what we normally mean by 'environment,' so if all can be privatized then we would, according to the marketeers, be well on the way to solving all environmental problems. We could go still further by privatizing species, wildlife, and fish. Species might be privatized through property rights to their genes. For example, rights to rare plants in endangered tropical forest ecosystems are claimed by pharmaceutical companies for the sake of their actual or potential role in producing new drugs. Wildlife might be privatized in conjunction with land, or, when animals wander across property boundaries, tracked by means of radio collars. Anderson and Leal (1991: 34) suggest that whales should be converted to private property: 'Whales also can be "branded" by genetic prints and tracked by satellite.' Conservationists wanting to save the whales could buy them, as could whalers wanting to hunt them. The market would determine the most appropriate balance. But note that whalers would not hunt to extinction, for once they had private property rights they would have every incentive to invest in the health of the whale stock, just as farmers invest in the health of their animals.

A market zealot would insist that solutions to environmental problems begin and end with the establishment of private property rights. As Coase (1960) demonstrated for the case of pollution, it does not even matter who has the right, the polluter or the sufferer from pollution, so long as it is legally clear. For if there is a legally unrestricted right to pollute, then market solutions to pollution can be generated by the sufferers banding together and offering to pay the polluter to cut back on emissions. Depending on what the sufferers are willing to pay, and whether this is larger or smaller than the profit the polluter is making from the activity, cutback will or will not occur in a fashion optimal in market efficiency terms. The fact that nowhere in the world can we observe sufferers offering to pay polluters to stop polluting does not stop Coase's article being regarded as a classic by economists.[2] (Scandinavian governments offered to pay Poland to stop polluting their atmosphere, but this is governmental action, not citizen-sufferers approaching polluters with an offer.)

Those who believe that if it moves you should privatize it, and if it doesn't move you should privatize it, represent the radical fringe of economic rationalism. Its stronghold is in US-based think-tanks such as the Foundation for Research on Economics and the Environment (FREE) in Seattle, the Political Economy Research Centre in Bozeman, Montana, the

Pacific Research Institute in San Francisco, the Independent Institute in Oakland, California, the Cato Institute in Washington, DC, and to a lesser extent mainstream conservative operations such as the American Enterprise Institute. There is a British counterpart in the Institute of Economic Affairs (London) and an Australian one in the Tasman Institute (Melbourne) (see Beder, 2001, for a survey of such think-tanks). Even in the United States, the privatizers have had little impact on the content of public policy.

More influential have been economic rationalists who advocate not wholesale privatization and private property rights, but rather market-type mechanisms and economic incentives to induce environmentally appropriate behavior, and to these I now turn.

If you can't privatize it, market it anyway

The hard-line economic rationalist position is that private property rights in air and water need to be established and enforced, nothing more. However, given the substantial difficulties with this hardline position, economic rationalists have often turned to the next best thing: government-managed markets and, failing that, quasi-market incentives. The most popular proposals for managed markets in the environmental realm involve pollution rights. Government defines an airshed or watershed, determines the maximum level of pollution that should be allowed, divides that level into a number of rights, then auctions off those rights to the highest bidder. After the initial auction has been held, polluters can buy and sell rights from one another. Polluters for whom it is easy and cheap to reduce emissions will cut back rather than pay for pollution rights, whereas polluters for whom emissions reduction is expensive will purchase rights to pollute. Thus the government-specified level of abatement will be achieved in the most cost-effective manner. Environmentalists who believe abatement should be greater still can always purchase quotas themselves and leave them unused (for arguments in favor of tradeable quotas, see Anderson and Leal, 1991: 145–7; Mitchell and Simmons, 1994: 155–7; Yandle, 1993).

Tradeable quotas have been introduced to a limited extent in the United States, where in 1979 the federal EPA began by sponsoring the 'bubble' concept in a few localities. However, in practice bubbles cover only a particular plant, and so the emissions 'trades' occur only within a company (that is, allowing the company to decide in which part of the plant it can most cheaply reduce emissions, rather than have government regulators instruct the company in what standards and technologies to use in particular parts of the factory). Bubbles and related trading practices promoted by the EPA resulted in hardly any trades in emission rights across companies, at least of the kind sought by economic rationalists (see Hahn, 1995: 134–7).

The 1990 US Clean Air Act Amendments allow for emissions trading on a larger scale for sulfur dioxide from coal-burning power plants. Under this Act, pollution credits beginning in 1995 were granted to 110 of the country's dirtiest coal-burning power plants, representing between 30 and 50 per cent of the sulfur dioxide currently emitted by them. The Chicago Board of Trade held auctions for additional credits. But this is still a highly restricted initiative and a far cry indeed from the purist economic rationalist position on tradeable quotas; the same can be said for all real-world US experiments so far. On a yet larger scale, the 1987 Montreal Protocol for the protection of the ozone layer provided for trades between countries in quotas for the emission of chlorofluorocarbons. European countries have generally not been keen on tradeable pollution quotas, though the United Kingdom began experimenting with emissions trading for carbon dioxide in 2002.

Tradeable quotas can also be established in resources such as fish (Stavins, 2002). The quota would refer to an allowable catch for a particular fishery for a specified time. Some government agency is needed to establish the quotas (and change them in response to the changing health of the fishery), but once established the quotas can be bought and sold on the market. Australia pioneered this system for its southern bluefin tuna fishery, and by 2004 had schemes in place for twenty-one fisheries. Tradeable quotas were introduced in Alaska's Pacific halibut fishery in 1995. Fisheries throughout the world have been notoriously subject to the tragedy of the commons, resulting in overfishing, depletion, and overcapitalization as fishers rushed to beat their competitors to the catch. Tradeable quotas have been less widely used than other forms of regulation (such as restrictions on numbers of boats, fishing seasons, equipment, and total allowable catch for the fishery as a whole).

More widely adopted than tradeable quotas are quasi-market incentives using standards and charges for pollution control, or 'green taxes' as they are sometimes known. Government sets an ambient environmental standard (for example, level of carbon monoxide in urban air), and then imposes taxes or charges on the activities which threaten that standard. The taxes in question can be levied on the goods whose production causes pollution, or directly on the pollution itself. Examples of the former are quite rare, though the European Community applies a tax on cadmium batteries, and several years ago the UK government mooted a proposal to impose an environmental tax on walking boots, on the grounds of the damage caused by the boots to footpaths in National Parks and other scenic areas of Britain. Examples of levies on pollution itself include charges per kilogram of sulfur dioxide emitted by smokestacks, or per kilogram of BOD (biological oxygen demand) for organic pollutants in rivers.

The economic rationalist's argument for a regime of green taxes is that they leave discretion in the hands of the polluter in terms of how much to reduce pollution and what kind of technology to use. If the polluter chooses to pursue abatement, then it has every incentive to find the most cost-effective means. Polluters for whom abatement is expensive will prefer to pay the charge and continue to pollute. All polluters have an incentive to search for less-polluting methods of production, for that will always save money. Government should set the charge per unit of pollution at a level sufficient to induce the required degree of abatement (for arguments in favor of green taxes, see Anderson et al., 1977; Kneese and Schultze, 1975; Moran, 1995).

Green taxes are not especially popular at the federal level in the United States (though hundreds exist at state and local levels; see Beck et al., 1998). Presidential candidate Al Gore renounced his previous commitment to green taxes in his 2000 presidential campaign. The idea of green taxes captured policy discourse in Britain in the late 1980s and early 1990s, when an upsurge in environmentalism coincided with a national government committed to market values. Prime Minister Thatcher remembered that she had studied chemistry long ago at Oxford University, and so could recognize chemical pollution when she saw it. The key document was produced by the environmental economist David Pearce for the Department of the Environment in 1989, entitled *Blueprint for a Green Economy* (Pearce et al., 1989), which advocated a comprehensive regime of green taxes (see Pearce and Barbier, 2000 for an update). This report was followed by a 1990 government white paper entitled *This Common Inheritance*, which curiously relegated the Pearce recommendations to an appendix. In late 1992 the government announced that 'In future, there will be a general presumption in favour of economic instruments' (quoted in Jacobs, 1995: 114). This presumption was slow to influence policy content. Part of the problem in Britain is that the Treasury interprets green taxes in revenue-raising terms, and would want to set levels without reference to environmental departments of government. This worries industry, which foresees charges rising and falling, most likely rising, in response to government's revenue needs. And it worries environmentalists, for it gives government a vested interest in pollution, for the more pollution that occurs, the more revenues does government receive (see Jacobs, 1995: 124). Because it sees green taxes in revenue-raising terms, the Treasury has generally opposed the use of the tax system for environmental purposes. In 1996 a tax was introduced on solid waste destined for landfills. The most prominent green tax in the United Kingdom is the congestion charge levied on vehicles entering central London, introduced in 2003. The charge has successfully reduced both congestion and air pollution. Mayor Ken Livingstone staked his political future on the charge—and won re-election in 2004.

Other countries have made more progress in implementing green taxes, especially on water pollution. France, Germany, and the Netherlands make use of per-unit pollution charges in their repertoire of environmental policy instruments (see Andersen, 1994). In France, charges are used mainly as a revenue-raising device, and are not set high enough to affect the environmental behavior of polluters. In the Netherlands, charges are successful and widely supported by environmentalists. In Germany, green taxes play only a secondary role within a more traditional regulatory system. German municipalities retain substantial control over policy implementation, and so happily dump pollution downstream. In all three countries revenue raised is earmarked for projects to improve water quality. As Hahn (1995: 146–7) notes, 'charges and marketable permits schemes . . . are rarely, if ever, introduced in their textbook form.' The same might be said of most policies inspired by economic rationalism: the textbook explication is crystal clear, the real-world implementation murky.

One international environmental problem that has received attention from advocates of green taxes is global warming, caused mainly by the buildup of carbon dioxide in the atmosphere from the burning of fossil fuels. Denmark, Finland, the Netherlands, Norway, and Sweden were the first to introduce a carbon tax, levied per tonne of fossil fuel burned. Other European countries followed. In 2001 the United Kingdom introduced a climate change levy on fossil fuels burned by industries and government bodies with high carbon dioxide emissions, though the coverage was restricted and accompanied by a complex system of exemptions. In the United States the Clinton administration proposed an energy tax, but this failed in Congress.

Before leaving green taxes, it should be noted that radical free-market zealots oppose them on the grounds that such taxes require competent and benign action on the part of government administrators in, for example, setting and changing tax rates (see, for example, Mitchell and Simmons, 1994: 148). Such zealots, recall, believe that the real cause of environmental problems is inadequate or inappropriate government specification of private property rights, and until that situation is rectified, any other policy actions are useless or counterproductive, and that includes green taxes.

Green taxes levied on goods are designed to induce consumers to make purchases that are less environmentally damaging. An alternative market-based means to the same end is provision of information about the environmental impact of a good, to facilitate green consumerism. The idea of 'eco-labeling' goods began in Germany in 1977, though arguably the most successful scheme is 'Nordic Swan' certification that has operated in the Nordic countries since 1989. Eco-labeling now ranges from forest products (which can be certified as free from destruction of tropical hardwoods) to organic food. Critics of green consumerism point out that it does not affect the total quantity of goods consumed by individuals, and that it is an easy symbolic alternative to confronting the structural causes of ecological destruction (Maniates, 2001). Yet confronting consumption seriously could in principle have a massive impact, because the pattern of consumption drives most economic activity (Conca et al., 2001). The problem is that the individual choices of green consumers are no match for the forces of corporate capitalism pushing environmentally irresponsible consumption—including 'greenwashed' products.

Analysis of economic rationalism discourse

Basic entities whose existence is recognized or constructed

Economic rationalism's world is populated by economic actors. *Homo economicus* can appear as a consumer or producer; and if producers are organized into firms, the firm still behaves like an individual. Markets, prices, and property have real existence. At some level government exists too as something more than a collection of economic individuals. However, economic rationalist discourse is ambiguous and troubled on this point. Some economic rationalists treat government as staffed entirely by *homo economicus* individuals, all concerned only with their own material interest, exploiting the public for personal benefit. This is why they always prefer markets to politics (see for example Mitchell and Simmons, 1994). But even these die-hard economic rationalists require someone, somewhere to be steering the system in the public interest, otherwise who is going to enact the appropriate arrangement of private property rights they seek?

Notably missing from economic rationalism are citizens (of the sort populating democratic pragmatism). Also, environments do not exist in any strong sense. At most, 'the environment' is only a pathway for some human decisions to have effects on other people—for example, through pollution. The existence of ecosystems, let alone ecosystems that often defy understanding, cut across chunks of private property, and impose constraints on human activity, is not perceived. There is no such thing as wilderness, only wilderness experiences (that is, human perceptions of wilderness amenity). There is an odd affinity here with postmodernists, for whom 'nature' is a human social construction. This lack of recognition of nature is driven home by Anderson and Leal's (2001: 27) comment on migratory bison that stray from Yellowstone National Park onto cattle ranches, potentially spreading brucellosis: 'The migration of the Yellowstone bison is like other cases of pollution in which the actions of one party, in this case the National Park Service, affect another, in this case Montana cattle ranchers.' So the bison are not recognized as part of the Yellowstone ecosystem, but instead demeaned as pollutants, at most the medium for one set of people (National Park Service officials) to affect another set of people (ranchers).

Unlike Prometheans, economic rationalists recognize the existence of natural resources, which is why it is crucial to establish the right kinds of property rights to these resources. In further contrast to Prometheans, economic rationalists would not necessarily dismiss the existence of limits to human activity imposed by finite resources.

Assumptions about natural relationships

Economic rationalism assumes that the basic relationship across individuals and collective actors (such as firms) is competitive. The co-operative problem solving sought by democratic pragmatists is ruled out. Corresponding to its thoroughly ambiguous attitude toward the existence of government as anything more than an assemblage (or sometimes tool) of rational egoists out to plunder the public purse, economic rationalism is confused about the existence of hierarchy within government. Administrative rationalism, as seen in Chapter 4, happily accepts hierarchy based on expertise. When it comes down to it, economic rationalists have to do the same, because some experts must be in a position of authority to implement appropriate private property rights, or to design green taxes. Of course, the experts themselves must be economic rationalists; but they cannot be economic *actors*, for if they were they would devise schemes in their own personal interest, not in the public interest.

The other kind of hierarchy implicit in economic rationalism is between humans and the natural world. Economic rationalism is thoroughly anthropocentric: nature exists only to provide inputs to the socio-economic machine, to satisfy human wants and needs. The appropriate expertise to manipulate these environmental inputs is taken for granted. Once appropriate property rights and incentives are in place, individual actors have no problem in deploying expertise to produce good results for society as a whole.

Agents and their motives

The main agents for economic rationalists are *homo economicus* ones, motivated by material self-interest, and pursuing it rationally. But, as I have just noted, exemption is granted for a few agents in governmental positions, who are allowed to be motivated by concern for the public interest, albeit defined in economic rationalist terms. Of course, the governmental actors who populate horror stories are not allowed any such public interested motivation; they are treated as rational egoists, whose interaction produces all kinds of perverse outcomes. Missing from economic rationalism is any notion of active citizenship; economic rationalism abolishes citizenship. When I received a circular from the economic rationalist government of the state of Victoria, where I used to live, it was addressed 'Dear Customer.' There were no citizens in Victoria.

Key metaphors and other rhetorical devices

Like the Promethean discourse analyzed in Chapter 3, the basic metaphor of economic rationalism is mechanistic. The social world is treated as a machine whose products meet human needs and wants, which can be understood through reference to its components and their functions. Unlike Prometheans, economic rationalists believe the machine may need to be reassembled, through, for example, redefinitions of property rights. Once we get the property rights in order, the machine will work smoothly. Environmental resources are treated as inputs to the social machine, be they raw materials for production or amenities such as wilderness and clean air.

Economic rationalists are skilled rhetoricians. Intervention by government administrators in the environmental affairs of industry and commerce used to be known, accurately and simply enough, as 'regulation.' Economic rationalists oppose regulation, so succeeded in stigmatizing it as 'command as control.' Little command and control actually occurs in environmental administration; there is much more in the way of informal cooperative relationships between government officials and polluters (see Chapter 4). So as a description of the real world, the term is laughable; but as a rhetorical ploy, it is brilliant. Following the collapse of Soviet-style systems which really did work by command and control, who could possibly favor such a system (except perhaps the military)?

An equally clever rhetorical ploy involves use of the adjective 'free,' especially to describe markets. A market is a market is a market; so why does it need to be called a free market, especially given that markets can only operate if government supplies a supportive legal context? Relatedly, why are capitalist corporations styled free enterprise? The answer is that the standard set of freedoms in liberal democratic societies is very popular.

In free markets and free enterprise, coercion is abolished if not in fact, then in rhetoric.

A third pervasive rhetorical strategy in economic rationalism is the horror story involving governmental action that produces perverse, inefficient, and costly results (for good selections, see Nelson, 1993; Stroup and Shaw, 1993). In the United States, one of the most widely circulated environmental horror stories picks up on the analysis of the 1977 Clean Air Act Amendments carried out by Ackerman and Hassler (1981) (who ironically are not themselves economic rationalists). Ackerman and Hassler demonstrate the disastrous results of a particular episode in legislation for environmental regulation. Eastern producers of high-sulfur coal combined with environmentalists to persuade Congress to mandate that all new coal-burning power plants install scrubbers to remove sulfur dioxide from their emissions, irrespective of how low the sulfur content of the coal was being burned, and so how much sulfur dioxide was being emitted. This measure effectively discriminated against western low-sulfur coal producers, and ensured that the ambient air quality targets would be met at a cost billions of dollars greater than could have been achieved with a switch to low-sulfur coal. Moreover, the legislation allowed existing coal-burning plants to operate with no additional controls, thus ensuring that old and dirty plants would gain a competitive edge over new plants, and so stay in use longer, thus actually encouraging increased pollution.

Other good horror stories include the Superfund, which shuffles toxic waste around at enormous expense to little benefit (Stroup and Meiners, 2000); water policy that, bizarrely, promotes water-intensive crops such as rice and cotton in the desert; and timber policy, which heavily subsidizes otherwise uneconomic logging. Most of these horror stories are true. Their rhetorical weight comes with the economic rationalists' generalization from these stories, nearly all of which are about the United States government, to all the environmental activities of all governments. Perhaps the fault lies with the US federal government in particular, rather than government in general.

BOX 6.1 Discourse analysis of economic rationalism

1. **Basic entities recognized or constructed**
   - Homo economicus
   - Markets
   - Prices
   - Property
   - Governments (not citizens)
2. **Assumptions about Natural Relationships**
   - Competition
   - Hierarchy based on expertise
   - Subordination of nature
3. **Agents and their Motives**
   - Homo economicus: self-interested
   - Some government officials must be motivated by public interest
4. **Key Metaphors and other Rhetorical Devices**
   - Mechanistic
   - Stigmatizing regulation as 'command and control'
   - Connection with freedom
   - Horror stories

An assessment of economic rationalism

Economic rationalism in environmental affairs has been around a long time now. Analysis and advocacy of quasi-market incentive systems has been the staple of environmental economics since the 1960s, and more radical market-oriented arguments gained ground in the 1980s. (So it is odd that such old policies have recently been recruited to the category of 'new environmental policy instruments'; see Jordan et al., 2003.) These arguments were quite consistent with the dominant political discourse of the 1980s, at least in the Anglo-American world, which since then has expanded to dominate international economic affairs. The US delegation at the 1997 Kyoto negotiations on climate change pushed international tradeable quotas in carbon dioxide emissions, yet ironically the United States has been unable to implement such quotas at home for pollution in general. International bodies such as the OECD and European Environment Agency have been pushing such policies for a long time. Yet the pace of diffusion of economic rationalism into environmental policy practice has been glacial. Regulatory policy instruments still dominate anti-pollution policy everywhere, tradeable quotas in access to resources are still rare, and there has been little privatization of resources and environmental goods. Even when policy instruments are adopted, there is no wholesale institutional change such as that sought by economic rationalists. Regulatory agencies are still with us and, far from being replaced by regimes of property rights or by economistic calculators of green taxes, they end up administering such schemes. When it comes to institutions, economic rationalism has had much less impact than administrative rationalism and democratic pragmatism which, as we saw in the two previous chapters, have been major factors in the evolution of environmental institutions in the last forty years. Economic rationalism's lack of influence at the policy level may reflect its failure at the institutional level, for the policy prescriptions have not been able to find a hospitable institutional home. Less commodious homes provide only for piecemeal and distorted adoption of economic policy advice.

No doubt part of the explanation for the glacial progress of economic rationalism in environmental affairs lies in simple inertia, and the resistance of established routines. Yet if inertia is so powerful, why has economic rationalism made greater inroads in areas such as labor-market deregulation, business deregulation, privatization of state-owned industries and utilities, even redesign of the welfare state? Another explanation might note that proposals for economic instruments can never enter in the clean and straightforward fashion of the economics textbooks. Instead, their entry and so their design is heavily dependent on the configuration of political forces and the prevailing political-economic context. We have seen in the case of Britain, for example, how the Treasury has treated green taxes in terms of their revenue-raising potential, but in doing so has made both industry and environmentalists nervous. More generally, businesses may oppose green taxes because they have to pay both abatement costs and the tax itself (Daugbjerg and Svendsen, 2003). So long as green taxes and tradeable permits are applied only sporadically, firms may not reorganize themselves to respond to the cost-saving incentives these instruments offer. In addition, regulatory standards may have been designed with input from existing firms in order to discriminate against newcomers—perhaps specifying more stringent standards for new pollution sources—and so existing firms may oppose market-based alternatives (Stavins, 2002).

The fairyland of neoclassical microeconomics in which economic rationalist argument for market-oriented policy instruments is rooted is very different from the real world. The good fairies are not in charge of policy design and implementation. Stavins (2002) points out that in the United States, environmental laws are written by a Congress full of lawyers who do not understand economics, and who are attracted by the symbolic politics that can please all sides by combining stringent standards with lax enforcement. But the environmental area is hardly unique in all these respects.

To get at deeper reasons for resistance to economic rationalism we need to treat it as a discourse rather than just a set of proposals for policies and institutions. Recall that the basic agents and motives recognized by economic rationalism treat people only as *homo economicus* consumers and producers. There are no citizens in economic rationalism. Dobson (2004: 1–5) suggests that economic incentives alone are unlikely to yield the substantive and multifaceted changes in behavior that a sustainable society requires. More seriously, such incentives may actually undercut ecological citizenship. Sagoff (1988) argues that all individuals have both consumer preferences and citizen preferences, and that these point in different directions. As a consumer, I may want to make use of freeways to get to work more quickly; as a citizen, I may demonstrate against construction of the freeways because they destroy communities and natural areas. We would normally put our citizen preferences first, though only if given the chance to express them in political settings. But economic rationalists count only consumer preferences, repressing citizen preferences. Theirs is a world unlikely to please environmental citizen activists, which may be why environmentalists have often opposed economic rationalist schemes. When we visit a national park, we do so as citizens. Part of the experience of being there is that it is indeed a *national* park, emblematic of what it means to be a Canadian, an American, a Costa Rican, or a Japanese; a repository of common trust and community pride. And visiting national parks in other nations can also involve recognition of and respect for the identity and citizenship of others. These are experiences that Walt Disney could never provide.

Opposition may also arise as a result of the way economic rationalism treats or, rather, does not treat the environment. Recall that in economic rationalist discourse, the environment exists only as a medium for the effects of some human actions on other humans, and as a source of inputs for the socio-economic machine. It is in this light that Anderson and Leal (2001: 27) can refer to migrating buffalo as 'pollution' because they may carry diseases that affect cattle. Thus the environment has no intrinsic value, and chunks of it can be bought or sold at will, depending only on the most profitable human use. When it comes to pollution, economic rationalism attaches no stigma: rights to pollute are just like any other commodity, to be bought or sold. As Kelman (1981) notes, this failure to stigmatize pollution in moral terms makes many environmentalists uneasy. Goodin (1994) compares this selling of pieces of the environment by governments to the selling of indulgences by the medieval Catholic Church. In both cases, individuals can have their sins forgiven if they can afford to pay. But just as places in heaven were not the church's to sell (only for God or St Peter to decide), so pieces of the environment are not government's to sell. Martin Luther and opponents of green taxes have more in common than they might think.

In short, no matter how attractive economic prescriptions may be in instrumental terms, even to committed environmentalists, they help constitute a discourse, and a world, which those according higher priority to citizenship, democratic, and ecological values find unattractive (see Dryzek, 1995). This is especially the case when prescriptions are tied to a general right-wing market agenda, as they are in the United States by 'free market environmentalists,' who thereby undermine the political prospects for the policies they favor. Paradoxically, some of these policies have been more easily implemented in social democratic North European countries, precisely because they do not attract right-wing ideological baggage there.

A further limitation of economic rationalism arises from its basically mechanistic metaphorical structure. The idea that the world may be full of complex ecological and social systems interacting in variable and uncertain ways is implicitly denied by economic rationalism. Economic rationalists have no way to deal with such interactions, which may violate the boundaries of private property rights, no matter how carefully drawn. For example, proposals for tradeable quotas in ocean fisheries inevitably treat species in isolation. But rational management of a single species is impossible. Whether the species survives or flourishes depends not just on how many tons of it are caught per year, but on what is happening to other species that are predators, prey, or competitors for the same ecological niche. Moreover, other factors such as pollution or development may affect the habitat of the species.

Finally, economic rationalism as a discourse gets into all kinds of tangles in its treatment of government. Its attitude to government is thoroughly ambiguous: at one level government is populated by rational egoists feeding at the public trough, plundering the public purse, and thoroughly indifferent to environmental values. At another level public-spirited government action is needed in order to put economic rationalist prescriptions into institutional and policy practice, and so economic rationalism depends crucially on administrative rationalism. The 'public choice' school of economic rationalists has thrived on horror stories about government in theory and practice; in fact, without quite realizing it, the public choice school has demonstrated that political order is impossible if everyone is a rational egoist (see Dryzek, 1992*b*). When it comes to environmental affairs, if everyone is a rational egoist, then the commons will always be abused, polluters will continue to generate externalities, and government will do absolutely nothing to remedy the situation. The obvious inference is that economic rationalism is inadequate as an orientation to environmental affairs (see Dryzek, 1996*b*). In this light, economic rationalism's real usefulness may come in detailing very precisely the destructive effects of *homo economicus*, and the need for his or her proclivities to be controlled by more socially, politically, and ecologically benign human motivations. Economic rationalism, unlike democratic pragmatism and green radicalism, has had nothing to say about these alternative wellsprings of human action.

This concludes my discussion of the three discourses of environmental problem solving. While all three have their problems, it is fair to say that the real-world achievements of administrative rationalism and democratic pragmatism are more substantial than those of economic rationalism. This conclusion will not necessarily dismay die-hard economic rationalists, who argue that the problem is precisely that their proposals have not been tested fairly. Still, for administrative rationalism and democratic pragmatism, their problems are revealed through examination of several decades of real-world influence. For economic rationalism, in contrast, problems are revealed by contemplation of the reasons for lack of impact. Partisans of each one of these three discourses often make their case through reference to the deficiencies of the other two, while remaining within the basic parameters of problem solving within the political-economic status quo of liberal capitalism. But the manifest difficulties of all three discourses lead others to be a bit more creative in looking for alternatives. Let me turn now to emerging discourses which remain reformist in their orientation to industrialism, but are more imaginative in seeking to dissolve familiar dilemmas and impasses.

NOTES

1 **www.epa.gov/adminweb/leavitt/enlibra.htm**.

2 There are good economic reasons why they do not. As Mancur Olson (1965) pointed out in his classic analysis of the logic of collective action, the fact that individuals share an interest does not mean they will act upon it. Each person has an incentive to take a 'free ride' on the efforts of others. This logic parallels that of the tragedy of the commons introduced in Chapter 2, in that rational individual decisions lead to collectively bad outcomes.

PART IV

THE QUEST FOR SUSTAINABILITY

The apocalyptic horizons of environmental concern were set in the early 1970s by survivalists who argued that economic growth and population expansion would have to yield to global environmental limits, sooner rather than later. Prometheans denied the existence of limits. The problem-solving discourses surveyed in Part III are essentially agnostic about global limits, focusing instead on the work to be done in the here and now. Yet problem solving is energized by the need to achieve some kind of resolution to conflicts between ecological values and economic values.

Life would certainly be less troublesome if such conflicts did not exist, or, failing that, could be dissolved. The unresolved dispute between survivalists and Prometheans could be put behind us, and environmental problem solving could proceed with renewed vigor in the knowledge that solutions are available that can respond effectively to a range of key ecological and economic concerns. Throw in commitments to global justice through the eradication of poverty and to the wellbeing of future generations, and the prospect would surely be irresistible. But what could possibly combine ecological protection, economic growth, social justice, and intergenerational equity, not just locally and immediately, but globally and in perpetuity? The answer is sustainable development, which specifies that we can have them all.

Since the early 1980s, sustainable development has become hugely popular as an integrating discourse covering environmental issues from the local to the global, as well as a host of economic and development concerns. Just what sustainable development means in practice is a matter of some dispute, as is the question of whether it can actually deliver on some, most, or all of its promises.

The notion of sustainability receives greater precision in the second discourse covered in Part IV: ecological modernization. Ecological modernization addresses the restructuring of the capitalist political economy along more environmentally defensible lines. The key is that there is money to be made in this restructuring. At one level ecological modernization is about the search for green production technology. But this search also opens the door to intriguing possibilities for more thoroughgoing transformation, involving political change as well as technological change. So although at first sight ecological modernization looks like a rescue mission for industrial society, albeit an imaginative one, it also points to political and economic possibilities beyond industrial society.

7

Environmentally Benign Growth: Sustainable Development

What is sustainable development?

Sustainable development refers not to any accomplishment, still less to a precise set of structures and measures to achieve collectively desirable outcomes. Rather, it is a discourse. Since the publication of the report of the Brundtland Commission in 1987 (World Commission on Environment and Development, 1987), it is arguably the dominant global discourse of ecological concern. As Torgerson (1995: 10) puts it, 'public discussion concerning the environment has become primarily a discourse of sustainability.' But just what is sustainable development? The most widely quoted definition is Brundtland's: 'Humanity has the ability to make development sustainable—to ensure that it meets the needs of the present without compromising the ability of future generations to meet their own needs' (World Commission on Environment and Development, 1987: 8). Later in the report Brundtland declares that 'In essence, sustainable development is a process of change in which the exploitation of resources, the direction of investments, the orientation of technological development, and institutional change are all in harmony and enhance both current and future potential to meet human needs and aspirations' (p. 46).

Sustainable development as a concept did not begin with Brundtland. The two words have been joined occasionally since the early 1970s, when sustainable development was actually a radical discourse for the Third World. The concept has a deeper history in the renewable resource management concept of maximum sustainable yield. The latter is the maximum catch from a fishery, or cut from a forest, or kill of game animals, that can be sustained indefinitely. But the maximum sustainable yield concept says nothing about growth in resource use (indeed, rules out growth), or about how management of different resources might interact, or what to do with non-renewable resources. Sustainable development is a much more ambitious concept in that it refers to the ensemble of life-support systems, and seeks perpetual growth in the sum of human needs that might be satisfied not through simple resource garnering, but rather through intelligent operation of natural systems and human systems in combination.

Brundtland's definition did not satisfy everyone, and other definitions of sustainable development proliferated. Opinions differ as to what human needs count, what is to be sustained, for how long, for whom, and in what terms. Attempts to take an analytical razor to the concept (e.g., Dobson, 1998) are only partially successful because they soon leave the ambiguities of the real-world discourse behind. In the early 1990s the Transportation Research Board of the United States National Academy of Sciences spent a million dollars trying to come up with a definition, but failed to do anything more than simply agglomerate the concerns of its members. By 1996 the United Nations Educational, Scientific, and Cultural Organization (UNESCO) was sponsoring a project to clarify the meaning of the concept in a number of disciplines, with a view to making the concept a scientifically usable one—implying that it was not yet a scientific concept.[1] Yet the UNESCO project has a difficult task. For the proliferation of definitions is not just a matter of analysts trying to add conceptual precision. It is also an issue of different interests with different substantive concerns trying to stake their claims in the sustainable development territory. For if sustainable development is indeed emerging as a dominant discourse, astute actors recognize that its terms should be cast in terms favorable to them. So environmentalists might try to build in a respect for intrinsic values in nature that is conspicuously missing in Brundtland. Third World advocates would stress the need for global redistribution, and highlight the needs of the poor to which Brundtland pointed. Business groups equate development with economic growth, such that sustainable development mainly means continued economic growth. Partisans of the limits discourse re-cast their survivalism in the language of sustainability. After endorsing Brundtland, those arch-survivalists Meadows et al. (1992: 209) go on to say that 'From a systems point of view a sustainable society is one that has in place informational, social, and institutional mechanisms to keep in check the positive feedback loops that cause exponential population and capital growth.' For Meadows and colleagues sustainability means an end to economic growth; for the World Business Council for Sustainable Development, sustainability requires perpetuation of economic growth. As the Council declares in its foundational document, 'Economic growth in all parts of the world is essential to improve the livelihoods of the poor, to sustain growing populations, and eventually to stabilize population levels' (Schmidheiny, 1992: xi).

Does this variety of meanings mean we should dismiss sustainable development as an empty vessel that can be filled with whatever one likes? Not at all. For it is not unusual for important concepts to be contested politically. Think, for example, of the word 'democracy,' which has at least as many meanings and definitions as does sustainable development. Part of what makes democracy interesting is this very contestation over its essence. Democracy is doubly interesting because just about everyone who matters in today's political world claims to believe in it. The parallels with sustainable development are quite precise. Just as democracy is the only game in town when it comes to political organization, so sustainable development became the main game (though not the only game) in environmental affairs, at least global ones. Sustainable development, like democracy, is a discourse rather than a concept which can or should be defined with any precision. The discourse itself does, though, have boundaries. Sustainable development is different from survivalism because while it recognizes that ecological limits should be respected, they can also be stretched if the right policies are chosen, so that economic growth can continue indefinitely. Langhelle (2000: 310–11) suggests that for Brundtland, at least, the limits in question were energy supply and climate change; though he also recognizes lingering ambiguities in the discourse on the question of limits. Sustainable development is different from Promethean discourse because it requires coordinated collective efforts to achieve goals, rather than relying on human spontaneity and ingenuity. And it is different from the varieties of environmental problem solving surveyed in the previous three chapters because it is much more imaginative in its reconceptualization of the terms of environmental dispute and in its dissolution of some long-standing conflicts.

The career of the concept

Prior to the 1980s, sustainable development was part of the environmentalist lexicon, especially in the context of discussions of developing societies in the Third World. The concept was explored as an alternative to mainstream interpretations of development as economic growth, which had failed to deliver. Impetus was received through contention by the emerging limits discourse that the Earth could not withstand a Third World that duplicated Western levels of affluence (Carruthers, 2001). Advocates were interested in the potential of appropriate technologies or intermediate technologies, which were low-cost, low in the environmental stress they imposed, and consistent with local cultural norms (see Schumacher, 1973). They preferred energy generation from cattle dung over nuclear power stations or large dams, small workshops over large factories.

The concept's prominence grew, and its meaning began to change, in 1980 with the publication of the International Union for the Conservation of Nature's *World Conservation Strategy*. But the real transformation into the contemporary discourse of sustainable development can be dated to 1983, when Gro Harlem Brundtland, Prime Minister of Norway, was asked by the Secretary-General of the United Nations to chair an inquiry into interrelated global problems of environment and development. Brundtland's World Commission on Environment and Development published its report, *Our Common Future*, in 1987. The report contains analyses and recommendations pertaining to the international economy, population, food, energy, manufacturing, cities, and institutional change. Its main accomplishment was to combine systematically a number of issues that have often been treated in isolation, or at least as competitors: development, global environmental issues, population, peace and security, and social justice both within and across generations. Brundtland developed a vision of the simultaneous and mutually reinforcing pursuit of economic growth, environmental improvement, population stabilization, peace, and global equity, which could be maintained in the long term. Such a vision was seductive, though Brundtland did not demonstrate its feasibility, or indicate the practical steps that would be required.

Since 1987 the discourse of sustainable development has flourished at the international level, especially inasmuch as international society is constituted by international governmental organizations (IGOs) and non-governmental organizations (NGOs). The Earth Summit, more formally the United Nations Conference on Environment and Development (UNCED), held in Rio de Janeiro in 1992, was a high point. The 171 national government delegations, many with heads of government present, gave sustainable development their stamps of approval (though the various delegations may have held to different meanings of the term). The Earth Summit endorsed *Agenda 21*, a lengthy and detailed follow-up to Brundtland's efforts, which argued that global environmental problems had arisen mainly as a result of the profligate consumption and production of the richer countries, but also recommended more economic growth for all to finance solutions. After the conference the United Nations established a Commission on Sustainable Development to implement *Agenda 21*, with special reference to how national and local governments might act, and how conflicts between First World and Third World notions of development and environmental protection might be resolved. Sustainable development advanced as a discourse for all, North and South, rich and poor; though the rich eventually lost sight of the global equity aspect that was central to Brundtland and her more radical predecessors (Meadowcroft, 2000: 379).

In 2002 Johannesburg hosted the World Summit on Sustainable Development (WSSD), the world's largest-ever international conference. The sheer number and variety of meetings held at the WSSD makes assessment difficult. The WSSD endorsed a 'Plan of Implementation' for *Agenda 21*. The plan was a little short on concrete measures, how they should be accomplished, and who exactly should do it (von Frantzius, 2004: 470), with the partial exception of targets and dates for improved access to clean water and sanitation for the world's poor. Thus sustainable development remained very much a discourse, rather than a plan of action. The WSSD saw some major repositioning in relation to the discourse. Wealthy states, long the champions of environmental concern at such gatherings, now seemed more interested in pushing the benefits of development that could be achieved through globalization and free trade (this was somewhat less true for the EU than the United States). And Third World governments, once skeptical about environmental concern as a luxury for the rich, now recognized the severity of their own environmental problems (Wapner, 2003: 4–6). Perhaps the most successful discursive repositioning was accompanied by the corporations present, which confirmed the status of business as a major participant in sustainable development, not a source of problems to be overcome. This role was solidified in partnerships involving business, governments, and NGOs, several hundred of which were established at the WSSD.

Outside summits, sustainable development has infused the discourse of international institutions. Even the World Bank, long castigated by environmentalists for its complicity in ecologically disastrous development projects (such as large dams and high-technology agriculture), has tried to improve its environmental image by establishing an Environment Department, appointing a Vice-President for Sustainable Development, and sponsoring a series of publications on environmentally sustainable development. The main theme of the Bank's 1992 *World Environment Report* was that environmental management and economic development could proceed together. Its 2002 *World Development Report* was organized around the idea of sustainable development, though it lost sight of the global equity aspect of the discourse, recommending that the rich countries could best help the poor by becoming still richer and providing bigger markets for poor countries' products. The Bank has also sponsored research on the development of indicators of sustainable development as alternatives to more established measures of national wellbeing such as gross national product (see, for example, World Bank, 1995). The EU has incorporated sustainable development in some of its constituent treaties, and saw the WSSD as an opportunity to distinguish itself from the more skeptical position of US negotiators. The EU proved the lone champion of renewable energy against the United States and Third World countries pushing expanded fossil fuel use (von Frantzius, 2004: 472).

While the sustainability discourse is most evident at this international level, it has made inroads within states (see Meadowcroft, 2000 for a catalogue). In 1990 Japan established a sustainable development program, with an eye to maximizing Japanese opportunities in the emerging sustainable eco-economy (opportunities which are not hurt by the existing energy-efficiency of the Japanese economy). In Brundtland's own Norway, ProSus (Program for Research and Documentation for a Sustainable Society) is a think-tank committed to sustainable development. In Australia, the federal government in 1990 set up an ecologically sustainable development process, with working groups on agriculture, energy, fisheries, forestry, manufacturing, mining, transport, and tourism. Symbolizing sustainable development's positive-sum approach to economy and environment, each working group contained representatives of both industry and environmental groups (along with government and trade-union officials). The working groups reported in 1992, and their efforts were incorporated into a National Ecologically Sustainable Development Strategy, though for domestic political reasons the process and the strategy subsequently languished (see Christoff, 1995).

In the United States, the sustainable development torch was carried in the Clinton administration by the President's Council on Sustainable Development, which could draw support from Vice-President Al Gore's personal views (Gore, 1992). However, the dominant US approach to sustainable development is captured succinctly by Bryner (2000): 'Sorry, not our problem', with little support in Congress, and no resonance for any broader public. With the exception of the United States, sustainable development received at least lip service from governments in the developed world (Lafferty and Meadowcroft, 2000), though none has addressed their own over-consumption of resources and stress on global ecosystems (Meadowcroft, 2000: 374). In Britain, the government initially endorsed Brundtland's stress on sustainable development but—astonishingly—asserted that existing British economic policy met these principles, further proof of just how far the concept can be stretched (Department of the Environment, 1988; see also Jacobs, 1991: 59). The British government's subsequent approach to sustainable development was halting and begrudging, though after 1997 Tony Blair's Labour government did set up a Sustainable Development Unit to examine the practices of all government departments (to little effect).

Among developed countries, sustainable development has been taken most seriously in Northern Europe. Researchers developing an Environmental Sustainability Index for the World Economic Forum rated Finland as the most sustainable country, followed closely by Norway and Sweden (as of 2002).[2]

International business is increasingly prominent. The International Chamber of Commerce and World Business Council for Sustainable Development, chaired by Stephan Schmidheiny of the Swiss company UNOTEC, were active at the 1992 Earth Summit. The Business Council was formed in 1990 at the invitation of Maurice Strong, secretary-general of the Summit. The Council is committed to economic growth, but with an environmentally sensitive face. Its component corporations such as 3M, Rio Tinto, Du Pont, Shell, Mitsubishi, and ALCOA can point to success stories in their own operations of environmentally aware practices such as recycling, efficiency benefits achieved by waste reduction, sustainable forestry, and energy-efficient production (see Holliday et al., 2002 for a compilation). By 2002 the Council was composed of 162 of the world's largest corporations, mostly from the manufacturing, mining, and energy sectors (membership is by invitation only). It was chaired by Philip Watts of Royal Dutch Shell. Not all of these 130 companies have exemplary environmental records. They included Enron, the energy supply corporation linked to President George W. Bush, before it went bankrupt in 2002.

Under the banner of 'Business Action for Sustainable Development', the Council was highly visible at the 2002 WSSD, where it mounted a concerted effort to publicize and embed the business view. The major statement launched at the WSSD by Holliday et al. (2002) argued that economic growth produced by free trade was the only hope for the world's poor. However, the Council did not propose growth at all costs, proclaimed commitment to corporate social responsibility, and joined with Greenpeace to criticize the United States' withdrawal from the Kyoto Protocol on climate change. The Council succeeded in establishing partnerships with business as the dominant tool for pursuing sustainable development. Cynics saw this as 'the privatization of sustainable development' (von Frantzius, 2004: 469), threatening to reduce the discourse to a series of commercial projects (Wapner, 2003: 4).

Where are the environmentalists in these developments? After all, sustainable development began life long ago as one of their concepts. Environmental groups have become less visible with time. But some environmentalists, such as Friends of the Earth Europe, have tried to keep up with the discourse, to remind everyone that sustainable development requires wholesale reductions in the stress that economic activity imposes on the environment, and respect for intrinsic values in nature. Environmentalists were present in large numbers at the WSSD, but their impact was much less obvious than that of business.

Discourse analysis of sustainable development

The core storyline of sustainable development once began with recognition that the legitimate developmental aspirations of the world's peoples cannot be met by all countries following the growth path already taken by the industrialized countries, for such action would over-burden the world's ecosystems. Yet economic growth is necessary to satisfy the legitimate needs of the world's poor. The alleviation of poverty will ameliorate what is one of the basic causes of environmental degradation, for poor people are forced to abuse their local environment just to survive. Economic growth should therefore be promoted, but guided in ways that are both environmentally benign and socially just. Justice here refers not only to distribution within the present generation, but also across future generations. Sustainable development is not just a strategy for the future of developing societies, but also for industrialized societies, which must reduce the excessive stress their past economic growth has imposed upon the Earth.

Basic entities whose existence is recognized or constructed

Sustainable development's purview is global; its justification rests in present stresses imposed on global ecosystems. But unlike survivalism, it does not stay at that global level. Sustainability is an issue at regional and local levels too, for that is where solutions will have to be found (as made clear in *Local Agenda 21*, whose principles have been adopted by local governments around the world). Thus the basic entities stressed in sustainable development are nested systems, ranging from the global to the local. The systems in question are both social and biological. Natural systems are not separate from humanity: as Brundtland put it: 'The environment does not exist as a sphere separate from humans ambitions, actions, and needs . . . the "environment" is where we all live' (World Commission on Environment and Development, 1987: xi). The biological components of systems are treated with more respect than the brute matter that Prometheans see in nature. While survivalists see problems in terms of global limits and solutions in terms of global management, sustainable development takes a more disaggregated approach. Particular resources and systems can be used and developed more or less wisely, imposing more or less environmental stress.

The Brundtland report itself is a bit ambiguous on the existence of limits. A statement that 'Growth has no set limits in terms of population or resource use beyond which lies ecological disaster' in part because 'accumulation of knowledge and the development of technology can enhance the carrying capacity of the resource base' is followed immediately by a recognition that 'But ultimate limits there are' (World Commission on Environment and Development, 1987: 45). These ultimate limits too prove capable of being stretched by technology. As Brundtland herself later put the point, 'The commission found no absolute limits to growth. Limits are indeed imposed by the impact of present technologies and social organization on the biosphere, but we have the ingenuity to change' (quoted in Hardin, 1993: 205). Ecological constraints should be respected, but once this is done economic growth can proceed indefinitely. Some commentators have tried to resolve the ambiguity here by distinguishing between 'strong' and 'weak' versions of sustainable development, the former explicitly recognizing limits, the latter denying them (see Hay, 2002: 214–17). But any such resolution leaves most sustainable development discourse somewhere between the two poles; the zone of ambiguity is much larger than the polar regions.

When it comes to social systems, sustainable development now takes the capitalist economy pretty much for granted (this was not true in the more radical discourse of the 1970s). However, the structure of political systems is not taken as given. The reorientation in problem solving that sustainable development prescribes may require shifts in power between different levels to meet more effectively the challenge of sustainability. The frequent appeals to coordinated international action and grassroots participation suggest that these shifts would be away from the nation-state as presently constituted to both higher (transnational) and lower (local) levels of political organization, as well as sideways to partnerships with business. Networked governance as an alternative to top-down administration (see Chapters 4 and 5) fits well here.

Assumptions about natural relationships

The most important relationship regarded as if not exactly natural then at least attainable is the positive-sum one: economic growth, environmental protection, distributive justice, and long-term sustainability are mutually reinforcing. In the contemporary world of sustainable development there are few hierarchies recognized in human affairs. Instead, there is co-operation. However, there is a hierarchy which puts human beings above the natural world. In keeping with its integration of a range of agendas, sustainable development can take the protection of nature on board. For example, Brechin et al. (2003) argue that the basic needs of the world's poor can be met while protecting biodiversity in the ecosystems on which they depend. But for the most part sustainable development remains anthropocentric. It is sustainability of human populations and their well-being which is at issue, rather than that of nature. Relationships of competition are de-emphasized, though it exists in the background capitalist economy. Sustainable development is to be achieved through cooperative rather than competitive effort (witness the partnerships that dominated the 2002 WSSD), distancing the discourse from both economic rationalists and Prometheans.

Agents and their motives

Sustainable development's key agents are not the global managers of the survivalists or the experts with a managerial hierarchy at their disposal of the administrative rationalists. Instead, the relevant actors can exist at many levels, consistent with basic notions about the existence of nested social and biological systems. In practice, sustainable development marginally de-emphasizes national governments and state actors, though states are still needed to construct international agreements and work with NGOs and business. In the 1980s sustainable development was established as a discourse of international society, especially as that society is populated by IGOs (such as the United Nations and the World Bank) and NGOs (such as global environmental groups). There is a role for the grassroots too: the green radical slogan 'think globally, act locally' can be adopted here. The Earth Summit's *Agenda 21* called for more citizen participation in environment and development decisions. And corporations have clambered on board the bandwagon to show that business too can play a constructive role. Sustainable development is sometimes cast as a discourse of and for global civil society (see Conca, 1994; Lafferty, 1996; Wapner, 1996), defined in terms of political interaction not encompassed by the state. But more traditional areas of state action are not excluded.

Key metaphors and other rhetorical devices

Prometheans and economic rationalists alike rely on mechanistic metaphors, treating the world as a machine whose bits can be arranged to better meet human needs. In contrast, sustainable development's metaphorical structure is organic. Organisms grow and develop; so can societies. Growth here is not just physical maturation that happens automatically, for sustainable development also stipulates self-conscious improvement. As such, it is consistent with notions of personal human growth that stress education and growing awareness, enabling the individual to negotiate his or her social environment in more effective fashion. The image is of an increasingly sensitive, caring, and intelligent human being—only, of course, it is sensitive, caring, and intelligent political-economic systems which are at issue, and the environment to be negotiated is not just a social one, but also a natural one. Just as in models that portray human development in terms of lifetime learning, the growth in political-economic capacities is seen as open-ended. The difference is that individual humans eventually die, whereas for sustainable development growth in political-economic capacities can go on in perpetuity.

The discourse does respect nature—to a point. But nature is treated mainly as something that provides useful services to humans. The 'natural capital' metaphor is sometimes invoked (Dobson, 1998: 41–7; Sachs, 1999: 33). That is, nature's capital stock deserves respect and should be sustained because it is imperfectly substitutable by man-made capital. This way of thinking about nature is very economistic.

Sustainable development in its very name links itself to the idea of progress, and progress is one of the most powerful notions in the modern world. Whatever their other differences, Victorian industrialists, Marxists, social democrats, liberal democrats, and market liberals have all believed in the essential idea of history moving in the direction of social improvement. Sustainable development carries this idea into an environmental era.

Sustainable development also involves a rhetoric of reassurance. We *can* have it all: economic growth, environmental conservation, social justice; and not just for the moment, but in perpetuity. No painful changes are necessary. This rhetoric of reassurance is far from the images of doom and redemption found in survivalism, or the horror stories beloved of economic rationalists. Advocates of sustainable development are more likely to highlight local success stories of sustainability than they are to dwell on instances of unsustainability (Holliday et al., 2002; Schmidheiny, 1992: 181–333).

Whither sustainable development?

If we were to look for sustainable development, where would we find it? As discourse, there is a lot of it about. But can we identify any practices and policies inspired by, committed to, and achieving sustainable development?

This question may not be quite the right one to ask, if we conceptualize sustainable development as a discourse rather than a target. But the same

BOX 7.1 Discourse analysis of sustainable development

1. **Basic entities recognized or constructed**
   - Nested and networked social and ecological systems
   - Capitalist economy
   - Ambiguity concerning existence of limits
2. **Assumptions about natural relationships**
   - Cooperation
   - Nature subordinate
   - Economic growth, environmental protection, distributive justice, and long-term sustainability go together
3. **Agents and their motives**
   - Many agents at different levels, transnational and local as well as the state; motivated by the public good
4. **Key metaphors and other rhetorical devices**
   - Organic growth
   - Nature as natural capital
   - Connection to progress
   - Reassurance

might be said of 'democracy,' yet this does not stop political scientists producing comparative studies of the quality of different democracies. However, such comparisons are rough and contested. It is easy to conclude that (say) contemporary Canada is more democratic than Russia under the Tsars, very hard to say which of Canada and Japan is today more democratic, harder still to conclude that Canada is a true democracy. The same applies to sustainable development. The World Economic Forum (WEF) has ranked 142 countries according to a sustainability index, and Finland, Sweden, and Norway occupy the top three places. But Finland's number one ranking does not mean it has achieved an adequate level of sustainability, and environmental groups were quick to point out Finnish shortcomings in forest management. The WEF index is controversial, and really just compiles measures of environmental performance rather than sustainability per se. As such, it misses the 'development' part of the equation. So the fact that Finland is on top gives no guidance to poor countries that would like to emulate Finland. This question becomes especially perplexing in light of sustainable development's core storyline, which specifies that poor countries cannot follow the growth path already taken by wealthy countries such as Finland without over-stressing the world's ecosystems.

Thus it is better to think of sustainable development not as a path taken by countries such as Finland, but as at most a discourse that will inspire experimentation with what sustainable development can mean in practice. Sustainability, like democracy, is largely about social learning, involving decentralized, exploratory, and variable approaches to its pursuit. Sustainable development (unlike survivalism) can be a multilayered and multifaceted enterprise. Rather than try to impose a common definition replete with an associated set of precise goals (which is what survivalists and administrative rationalists would do), a 'decentered' approach would stress pluralistic and local experimentation (Brooks, 1992; Torgerson, 1994; 1995). In this search, the very fact that agreement on the essence of sustainable development has been elusive proves to be a help rather than a hindrance, for no avenues are ruled out by stipulation, and so all kinds of new possibilities might be unearthed (Torgerson, 1994: 310–13; see also Thompson, 1993).

But if the pursuit of sustainability is to be decentered and piecemeal, what would actually harness all these efforts to the common good? The answer lies in the necessity for widespread commitment to the discourse itself, the only conceivable glue to hold these various efforts together. In this light, the sought-after restructuring of power relationships becomes understandable. Sustainable development is a discourse of and for global civil society, not just states. Luke (1997: ch. 6) interprets this feature quite cynically as simply serving the interests of managerial 'ecocrats' employed in IGOs and NGOs. Luke's argument would be plausible if sustainable development did indeed constitute a unified approach. But with the decentered, piecemeal twist, the role played by global civil society can become democratic rather than managerial, an antidote to governments increasingly under the sway of market liberal ideas and committed to reducing environmental controls, expanding trade, and promoting economic growth at all costs (Lafferty, 1996). The problem is that market liberalism is now a powerful discourse in the international system itself, furthered by the same corporations now so active in the international politics of sustainable development.

There is no guarantee that widespread commitment to and pursuit of sustainable development in piecemeal fashion will deliver the goods. Economic rationalists see the whole enterprise as just the latest in a long line of futile attempts to replace markets by political management, trying to impose a discipline on people's decisions which is properly exercised by the market's price system (see Anderson and Leal, 1991: 167–71).[3] Prometheans see a lingering stress on limits poisoning the discourse (for example, Beckerman, 2002). Radical environmentalists deny that development (interpreted as economic growth) can ever be sustainable, and denounce the anthropocentric arrogance implicit in the discourse (for example, Merchant, 1992; Richardson, 1994). Radicals also argue that in an age of market liberalism, sustainable development's promise of social justice is hollow, as inequalities between rich and poor expanded in the 1990s within and across nations (Carruthers, 2001: 103). Even moderate environmentalists might wonder whether sustainable development diverts their energies by asking them to take on all the problems of the world, poverty and economic development as well as environmental protection (Wapner, 2003: 10). Survivalists attack any denial of limits and carrying capacity explicit in the discourse; so Garrett Hardin (1993: 204–6) takes Brundtland to task for failing even to ask whether the population growth she sees as inevitable and the economic growth she sees as desirable can be accommodated by the earth's resources (see also Milbrath, 1989: 320–3). Similarly, Herman Daly (1993) believes that Brundtland's vision of a world economy five to ten times larger than its present size is impossible given that the present human economy already appropriates 25 per cent of the world's 'net primary product of photosynthesis.' The more pessimistic conservation biologists argue that resources are rarely managed with sustainability in mind until after they have collapsed, for only then does their overuse become apparent (Ludwig et al., 1993; but for a catalogue of local cases where resources have been managed sustainably, see Ostrom, 1990).

Such criticisms notwithstanding, Lafferty (1996) argues that there is simply no better vehicle than sustainable development for environmentalists to pursue their various goals. The different strategic choices made by some eminent survivalists are noteworthy in this context. Meadows et al. (1992) disguise their survivalism in the words of sustainable development, and praise Brundtland; Hardin (1993) rubbishes sustainable development, and berates Brundtland.

The success or failure of sustainable development rests on dissemination and acceptance of the discourse at a variety of levels, followed by action on and experimentation with its tenets. Yet the twenty years that have seen sustainable development establish itself as the leading transnational discourse of environmental concern have seen much less in the way of wholesale movements in policies, practices, and institutions at global, regional, national, and local levels. Those same twenty years have seen a more effective global movement in a very different direction about which sustainable development is sometimes silent, sometimes (in its business-friendly variant) accepting. That direction involves the increasing transnationalization of capitalism, especially following establishment of the World Trade Organization in 1994. The WTO joined the International Monetary Fund and World Bank in policing international economic regimes. Free trade, capital mobility, and governments all over the world committed to market liberalization and ordinary (unsustainable) economic growth as their first imperatives threaten to override sustainable development. At the WSSD there were no serious suggestions that the WTO could be made to submit to sustainable development, but plenty of arguments from developed countries' national delegations that on trade issues the WSSD had to proceed in the context set by the WTO.

In a world dominated by market liberalism, sustainable development's prospects are poor unless it can be demonstrated that environmental conservation is obviously good for business profitability and economic growth everywhere, not just that these competing values can be reconciled. As we shall see in the next chapter, this is exactly the claim advanced by ecological modernization.

NOTES

1 This project was under the auspices of UNESCO's Management of Social Transformation Program, and organized through the Institute for Social-Ecological Research in Frankfurt, Germany.

2 The rankings are available online at **www.ciesin.org.indicators/ESI/rank.html**.

3 Anderson and Leal regard sustainable development as sufficiently important for its contrast with their free-market environmentalism to form the conclusion of their book, which is widely regarded as the definitive statement of economic rationalism applied to environmental affairs. However, they wrongly assert that sustainable development involves a globally administered regime of zero economic growth and zero use of non-renewable natural resources. In other words, they mistake sustainable development for an extreme form of survivalism.

8

Industrial Society and Beyond: Ecological Modernization

Cleanest and greenest

Mirror, mirror on the wall, who is the greenest of them all? Which countries turned in the most successful environmental policy performance in the 1980s and 1990s? Among developed nations, consensus picks include (in alphabetical order):

- Finland
- Germany
- Japan
- The Netherlands
- Norway
- Sweden

Of course, different indicators of environmental policy success produce different rankings, and some dimensions of environmental conservation (such as biodiversity protection) are not easily measured. So any such ranking is likely to be controversial, especially from the point of view of those not at the top of the list. The most comprehensive ranking of 142 countries according to 68 measures has been carried out by Yale and Columbia researchers working for the World Economic Forum (WEF).[1] The indicators range widely over air and water quality, number of threatened species, change in forest cover, infant mortality, renewable energy as percentage of total energy use, civil liberties, and democratic government (justified on the grounds of institutional capacity). The summary index placed Finland, Norway, and Sweden at the top. Focusing more narrowly on changes in pollution levels over time, Scruggs (2001) identifies Germany as the top performer across sixteen developed countries in the 1980s and 1990s. Emphasizing level of as well as change in pollution, Jahn (1998) has the Netherlands as the top performer.[2] Japan is not so clearly at the top of any league table, though it can claim leadership when it comes to energy efficiency (the amount of energy required to produce national income), and in its development of pollution control technology (Revell, 2003: 24–48). These six countries have been very supportive of international initiatives for environmental protection.

Comparative statistics tell only part of the story. If we dig a little deeper, we find that these six countries have adopted innovative and advanced procedures, policies, and institutions for dealing with environmental issues. Around 1970, environmental policy innovations mostly began in the United States, and were then copied elsewhere. But since the 1980s, the United States has fallen behind, stuck in a standoff between supporters and opponents of the laws and regulations established around 1970 (Bryner, 2000: 277). Let us take a look at what others have been doing in the meantime.

In 1989 the Netherlands adopted a National Environmental Policy Plan designed to integrate environmental criteria into the operations of all departments of government. A plan has been published every four years since then, and every year a State of the Environment Report ascertains progress. The plan is oriented by a set of environmental quality targets along with a timetable for achieving them, and grounded in a sophisticated theory of how pollutants are generated in and travel through human social systems. Rather than control pollutants at the end of the pipe, the Dutch plan seeks to identify and change activities that cause pollution in the first place. The changes are identified in consultation with the relevant industry, citizen groups, and responsible government officials, especially those from departments dealing with industry, agriculture, and transport. Under the plan, which relies on collaboration, not rules and penalties, 250,000 businesses have agreements with government. The plan encourages energy-efficient manufacturing and transport, agriculture that can achieve good yields while minimizing use of herbicides and pesticides, conservation of biodiversity, and so forth. The 2001 plan began to address the international dimension of Dutch environmental issues. None of this is done piecemeal, but rather in the context of the targets of the plan as a whole. The driving idea is that economic growth should be delinked from rising environmental stress, though this has yet to be achieved in practice. The environment is not treated as a policy area to be dealt with in isolation. Instead, environmental concerns are woven into all the relevant areas of government. The National Environmental Policy Plan has had its political ups and downs, and there have been inevitable disappointments in its implementation. However, as a process for 'turning government green' as Weale (1992: 122–53) puts it, the plan remains a landmark, though the Netherlands has not yet got as far as restructuring the economy on more ecological lines (van Muijen, 2000).

In Germany, concern over the implementation deficit associated with earlier environmental policies led in the 1980s to the adoption of the *vorsorgeprinzip*, or precautionary principle, as the guiding force for policy. The precautionary principle specifies that scientific uncertainty is no excuse for inaction on an environmental problem. Thus if there are good reasons for thinking a problem may be serious, it will be addressed, even in the absence of scientific proof. Dealing with a problem immediately and cheaply is better than waiting for it to get worse, by which time the amount of money required to fix the problem may have multiplied many times over.

The precautionary principle was strongly resisted—indeed, barely comprehended—by the national governments of Britain and the United States in the 1980s. In the US, the administrations of Reagan and Bush the elder used scientific uncertainty as an excuse for inaction on acid rain, especially over claims that sulfur emissions in the United States caused acid rain that damaged lakes and forests in Canada. This sort of excuse can also be found in the George W. Bush administration when it comes to climate change. In Britain, the absence of conclusive science was long the standard governmental excuse for inaction on every major regional and global pollution issue: acid rain, carbon dioxide, chlorofluorocarbons, coastal pollution, and sludge dumping in the North Sea, among others (matters began to change in the late 1990s). Germany moved ahead in tackling all the pollution problems that Britain denied. By the mid-1990s environmental protection was established as a goal in the German constitution (which in a legalistic state really matters), along with a comprehensive and complex set of environmental laws (Jänicke and Weidner, 1997).

Japan stands out largely because of the energy-efficiency of its economy. This efficiency may be explained to a degree by the extent to which Japan depends on imported oil, and so was shaken by the energy crises of the 1970s. We find in Japan environmental policy made with a minimum of fuss and a maximum of consensus. As in Japanese policy making on all major issues, the key players are government officials and business executives. Like the Netherlands, Japan has aimed to decouple economic growth from environmental stress in terms of pollution (Barrett and Fisher, 2005). Japanese politicians have recognized economic possibilities in the export of green technologies, together with 'green' public work projects as an alternative to the traditional approach to dispensing money to regions that involves covering Japan with concrete structures nobody needs.[3]

Norway has its environmental blemishes, most notably its continued support for commercial whaling. But, as befits the home of Gro Harlem Brundtland (the Sultana of Sustainable Development), Norway has made strenuous efforts to incorporate environmental values into policy making. It has pioneered policy instruments such as green taxes. The policy-making structure of Norway is corporatist in that economic and social policies are made behind closed doors by a small number of leaders from government, the labor-union federation, and the business sector. Norway, however, is unique in admitting environmental groups to the center of corporatist policy making. Thus the Norwegian Society for the Conservation of Nature, also known as Friends of the Earth Norway, is largely funded by government through operating grants, and is represented on key policy-making committees. The society also helps implement government policy through receiving project grants (Dryzek et al., 2003: 22–7). This situation is very different from that in the United States and Britain, where Friends of the Earth is a campaigning group that tries to influence government from the outside.

Sweden pioneered integrated pollution control (see Weale, 1992: 97–100). In most countries, anti-pollution policy is organized around single-medium and single-substance legislation and regulation. The result is that one pollutant may be reduced, but another pollutant increased as a result. For example, a pollutant discharged into a watercourse may be eliminated by collecting it as a toxic sludge, which might then be dried and burned, leading to air pollution. In Sweden, licenses for new manufacturing plants are issued only after a consideration of the total emissions of the plant, and what might be done to reduce that total to an acceptable level. Along with the Netherlands, Sweden has led in integrating environmental principles across all departments of government, coordinated by key cabinet ministers serving on a Delegation for Ecologically Sustainable Development (Lundqvist, 2004).

Finland, the top environmental performer according to the WEF Sustainability Index, adopted the world's first carbon tax in 1990. Despite the cold climate, carbon dioxide emissions are relatively low. Finland has pioneered other environmental policy instruments (Sairinen, 2003), and has made substantial progress in reducing pollution levels. Finnish industry sees environmental performance as a competitive advantage, so reforms can draw on consensus spanning industry and environmental groups.

What do Germany, Japan, the Netherlands, Norway, Sweden, and Finland have in common that might explain their apparently superior environmental performance? The first three are densely populated countries that have largely destroyed their native ecosystems (replaced by agro-ecosystems and urban ecosystems), and so have strong incentives to find a way to accommodate a dense population while minimizing further environmental damage. But the same might be said for environmental laggards like Britain, Belgium, and Denmark. And Norway, Sweden, and Finland have a relatively low population density (at least by European standards). The environmental movements are not any stronger in these countries than comparable others. Indeed, in Norway the movement's numbers are tiny compared to similar countries (Dryzek et al., 2003: 24). The Green Party has played a key role in German policy development, partly by forcing other parties to adopt green positions for fear of losing votes to the Greens. But the other five countries lack a green party of comparable force (though Greens have participated in the national government in Finland); Japan and Norway have no green party of any consequence.

What these countries have in common is a political-economic system where consensual relationships among key actors prevail. In discussing Norway, I introduced the idea of corporatism. They are all, to greater or lesser degrees, corporatist systems. Japan can be described as 'corporatism without labor,' leaving only government officials and business leaders to cooperate in policy formation (Lehmbruch, 1984). Thus the six countries all eschew both adversarial policy making and unbridled capitalist competition. Their polar opposites in these respects are the English-speaking developed countries: Britain, the United States, Canada, Australia, and New Zealand. Jahn (1998: 120) finds a clear positive relationship between degree of corporatism and environmental policy success. Until the 1970s, corporatist systems were all organized to emphasize issues of economic growth and income distribution. Yet once environmental values were taken on board, corporatism eventually enabled these values to be addressed in a particular fashion: that of ecological modernization. Here lies the key to apparently superior performance.

The idea of ecological modernization

Ecological modernization was first identified in the early 1980s by the German social scientists Joseph Huber (1982) and Martin Jänicke (1985), who observed and interpreted its development in Germany. Ecological modernization refers to a restructuring of the capitalist political economy along more environmentally sound lines. Environmental degradation is seen as a structural problem that can only be dealt with by attending to how the economy is organized, but not in a way that requires an altogether different kind of political-economic system (Hajer, 1995: 25). Environmental criteria must be built into the re-design of the system, as in the Dutch National Environmental Policy Plan.

Conscious and coordinated intervention is needed to bring the required changes about. It is no good relying on any supposed 'invisible hand' operating in market systems to promote good environmental outcomes (of the sort Prometheans stress). Yet this intervention does not take place in adversarial fashion, in terms of government imposition. Industry itself cooperates in the design and implementation of policy. The key to ecological modernization is that there is money in it for business. Thus business has every incentive to embrace rather than resist ecological modernization, provided only that business is sufficiently far-sighted, rather than interested only in quick profits.

What exactly is in it for business? First, 'pollution prevention pays,' as a popular slogan has it. Pollution is a sign of waste. Less pollution means more efficient production. Second, if a problem is not solved in the present, solving it in the future may be vastly more expensive for both business and government. For example, poorly managed toxic waste dumps become a stew of dangerous chemicals leaking into ground water, soil, and the air. To clean them up is extraordinarily expensive (as the experience of the Superfund in the United States demonstrates). Far better and far cheaper not to let such problems develop in the first place. Third, an unpolluted and aesthetically pleasing environment means healthier, happier, and more productive workers, who may even willingly sacrifice wages and salaries for these environmental rewards. Fourth, there is money to be made in selling green goods and services. Consumers increasingly demand products that are not excessively packaged, that do not contain artificial and toxic ingredients, and that are not produced in environmentally damaging ways. Fifth, there are profits to be had in making and selling pollution prevention and abatement products.

Traditionally, increased national income per head has gone hand-in-hand with increased stress on the environment. As an old Yorkshire saying has it, 'where there's muck there's brass.' Successful ecological modernization would decouple muck and brass, such that income per head could go on increasing without additional strain on the environment. This possibility would, it seems, dispel the darkest fears of the survivalists. A qualitatively different kind of growth would not have to hit ecological limits, even if those limits did have real existence.[4] Reconciliation with the overarching need for governments to promote economic growth means that environmental values now support economic ones, which in turn allows (moderate) environmentalists to be included in the core of policy making (Dryzek et al., 2003: 64–5). This inclusion has been most successful in corporatist political systems, with Norway the very best example.

Ecological modernization is sometimes treated as a merely technical concept, referring to the re-tooling of industry and agriculture along more environmentally sensitive but still profitable lines. Yet if this is the case, there is nothing truly 'ecological' about it, for it would say little about human interactions with ecosystems (see Christoff, 1996*a*). There really has to be more to the discourse than narrow engineering and technical concerns. For ecological modernization is not something that can be accomplished by business managers and engineers operating voluntarily and independently on their own products and processes. It requires political commitment, to the enlightened long term rather than the narrow-minded short term, and to a holistic analysis of economic and environmental processes rather than piecemeal focus on particular environmental abuses. Its subject matter encompasses nothing less than how capitalist society shall be guided into an environmentally enlightened era, and so involves commitments on the part of the entire society, not just industry. These commitments include foresight, attacking problems at their origins, holism, greater valuation of scarce nature, and the precautionary principle. There is a role for government in setting standards and providing incentives to industry, which helps explain why ecological modernization has flourished in countries with interventionist governments that work closely with business.

Ecological modernization bears a family resemblance to sustainable development. In his seminal book on the subject Hajer even categorizes the Brundtland report as a key ecological modernization document (Hajer, 1995: 26). But ecological modernization has a much sharper focus than does sustainable development on exactly what needs to be done with the capitalist political economy, especially within the confines of the developed nation-state.[5]

Discourse analysis of ecological modernization

The storyline of ecological modernization is that the capitalist political economy needs conscious reconfiguring and far-sighted action so that economic development and environmental protection can proceed hand-in-hand and reinforce one another.[6] This storyline is constructed from the following discourse elements.

Basic entities whose existence is recognized or constructed

Ecological modernization entails a systems approach that takes seriously the complex pathways by which consumption, production, resource depletion, and pollution are interrelated. This is most explicit in the Dutch National Environmental Policy Plan, which, as Weale (1992: 128) points out, is rooted in general systems theory. The key to effective action is therefore to anticipate and prevent unwanted environmental ramifications of production and consumption decisions. This orientation is very different from the atomistic underpinnings of Promethean and economic rationalist discourse, which have little time for system complexity. However, ecological modernization's embrace of the system concept is incomplete, for it can still view natural systems in limited terms, as mere adjuncts to the human economy. Nature is treated as a source of resources and a recycler of pollutants—a giant waste treatment plant, whose capacities and balance should not be overburdened. Denied are any notion that nature might spring surprises on us, defy human management, have its own intrinsic value, and its own open-ended developmental pathways. This limited view of nature warrants green radical suspicion of ecological modernization.

Like sustainable development, ecological modernization pushes limits to growth into the background. This displacement of the limits discourse is symbolized by the 1997 report to the Club of Rome, *Factor Four*, which argued the compatibility of doubling wealth while halving resource use (von Weizsäcker et al., 1997). In 1972 it was the Club of Rome that commissioned the key survivalist report, *The Limits to Growth*. In ecological modernization discourse, limits are not so much explicitly denied as ignored. Certainly the idea of limits becomes fuzzier once economic growth is decoupled from growth in environmental stress, which seems to be happening in the six countries I identified at the beginning of this chapter. The existence of the capitalist political economy is taken for granted. Unlike sustainable development, economic redirection does not necessarily require a de-emphasis of the state and concomitant promotion of international society and the political grassroots. Finland, Germany, Japan, the Netherlands, Norway, and Sweden are strong states which if anything become stronger as a result of their promotion of ecological modernization (but see Mol, 1996: 314–15 for an argument that ecological modernization can allow a more participatory and decentralized state).

Assumptions about natural relationships

Ecological modernization implies a partnership in which governments, businesses, moderate environmentalists, and scientists cooperate in the restructuring of the capitalist political economy along more environmentally defensible lines. This partnership is an anthropocentric one, in that the natural world is subordinate to human desires and calculations. Whether or not there are necessary hierarchies in human affairs is an open question. Certainly, there are those who would want to make ecological modernization into a doctrine for managers of the political economy; on the other hand, there is room for more egalitarian political relationships across different actors. There is also a crucial natural relationship between environmental protection and economic prosperity, in that the two are seen as properly proceeding hand-in-hand.

Agents and their motives

The key agents in ecological modernization are the partners I have just identified: governments, businesses, reform-oriented environmentalists, and scientists. Their motivations have to do with the common good or the public interest, defined in broad terms to encompass economic efficiency and environmental conservation. Ecological modernization requires widespread commitment to and action upon its principles. If it is resisted by key actors, as in the United States and United Kingdom, then ecological modernization simply will not happen.

The question of agency under ecological modernization does, on closer inspection, provide a doorway into a potentially more far-reaching change in the way developed societies organize their economic and—especially—political systems. For the partnership is in a major enterprise: the ecological restructuring of capitalism. But the partnership itself might prove to constitute a major restructuring of political life, because its scope will be extended to questions of economic organization that have traditionally been placed off-limits to collective political control. Shortly, I will turn to the radical ramifications of this possibility.

Key metaphors and other rhetorical devices

The words 'economics' and 'ecology' both derive from the Greek *oikos*, meaning household. In a sense, ecological modernization returns both ecology and economics to their household root, and re-establishes their commonality. For the implicit metaphor in ecological modernization, helping to explain its widespread appeal, is that of a tidy household. This household is concerned with maximizing its wellbeing, but at the same time realizes that minimizing waste also means meeting its needs efficiently, and that commodious surroundings contribute to the household's sense of wellbeing. In this light, perhaps it is not surprising that ecological modernization has prospered in countries noted for the tidiness, prudence, and far-sightedness of their households.

The word 'modernization,' like the word 'development,' connotes progress, and so ecological modernization is linked with the ever-popular notion of social progress. Again like sustainable development, ecological modernization is a discourse of reassurance, at least for residents of relatively prosperous developed societies. No tough choices need to be made between economic growth and environmental protection, or between the present and the long-term future. Unlike sustainable development, it is rarely claimed that this happy coincidence of values extends to social justice, still less justice across the rich and poor nations of the world. However, Hawken et al. (1999: 1–2) say that justice too will follow if the technological changes they advocate are adopted. Ecological modernization was long silent about what might be the appropriate developmental path for Third World societies, and attempts by theorists to apply ecological modernization concepts to these societies often misfire. To get to the point where they can now choose ecological modernization, countries like the cleaner and greener six spent a lot of time in a modernization mode that was decidedly anti-ecological. If followed by the world's poor, then that developmental path would surely impose intolerable stress on the world's ecosystems. Sustainable development speaks more explicitly to Third World development than does ecological modernization.

Radicalizing ecological modernization?

In its limited technical sense, ecological modernization looks like a discourse for engineers and accountants. However, ecological modernization can also be treated as a restructuring of political and economic life, rather than a mere re-tooling of industry.

At one extreme we find ecological modernization for engineers and accountants. Hajer (1995) refers in this context to 'techno-corporatist' ecological modernization, which treats the issues in technical terms, and seeks a managerial structure for their implementation. Management is

BOX 8.1 **Discourse analysis of ecological modernization**

1. **Basic entities recognized or constructed**
   - Complex systems
   - Nature as waste treatment plant
   - Capitalist economy
   - The state
2. **Assumptions about natural relationships**
   - Partnership encompassing government, business, environmentalists, scientists
   - Subordination of nature
   - Environmental protection and economic prosperity go together
3. **Agents and their motives**
   - Partners; motivated by public good
4. **Key metaphors and other rhetorical devices**
   - Tidy household
   - Connection to progress
   - Reassurance

supplied by the existing administrative organization of the corporatist state, open to the findings and recommendations of environmental scientists and engineers. Relatedly, Christoff (1996*a*) refers to 'weak' ecological modernization, characterized by:

- an emphasis on technological solutions to environmental problems;
- a technocratic/corporatist style of policy making monopolized by scientific, economic, and political elites;
- restriction of the analysis to privileged developed nations, who can use ecological modernization to consolidate their economic advantages and so distance themselves still further from the miserable economic and environmental conditions of the poorer nations of the world.

Christoff's 'strong' ecological modernization would feature in contrast:

- consideration of broad-ranging changes to society's institutional structure and economic system, with a view to making them more responsive to ecological concerns;
- open, democratic decision making maximizing not only participatory opportunities for citizens, but also authentic and competent communication about environmental affairs;

- concern with the international dimensions of environment and development.

An excellent illustration of the contrast between weak and strong ecological modernization, especially when it comes to the difference between technical and structural solutions, can be found in González's (2001*b*) analysis of air pollution policy in California. California has long pursued a policy of forcing technical changes to car engines in order to reduce emissions. However, total emissions continue to rise because the benefits of these changes are more than offset by increases in the number of cars on the road and average per-year distance traveled. Planning to reduce reliance on private cars and control urban sprawl is not on the agenda. Such planning would be central to strong ecological modernization.

Consistent with Christoff's 'strong' viewpoint, Hajer (1995) speaks of the possibility of 'reflexive' ecological modernization. By reflexive, Hajer means political and economic development that proceed on the basis of a critical self-awareness. Modernization was for long treated in nonreflexive terms as just a matter of hitching a ride on the ineluctable progress from 'traditional' to 'modern' society. Reflexive modernization still recognizes that the ride must be taken, but introduces a host of anxieties about the quality and trajectory of the ride which must be subject to continued monitoring and control. No longer can experts and governments be trusted to know what is best for the rest of us; no longer should we regard economic growth of whatever composition as automatically a good thing; no longer should we place economic affairs and the organization of the economic system as off-limits to public scrutiny and democratic control. Experts and elites would have to justify their policies in front of the citizens, in comprehensible language, and with no recourse to the privilege of rank or expertise. Reflexive ecological modernization is for everybody.

Clearly it matters a great deal to which of these two versions of ecological modernization a society commits itself. Weak or techno-corporatist ecological modernization might, as Hajer (1995: 32–4) recognizes, involve just a rhetorical rescue operation for a capitalist economy confounded by ecological crises. This would defuse the radical potential of environmentalism and deflect the energies of green activists without really changing the political-economic system to make it more ecologically sustainable and socially convivial.

Much more is at stake in the strong version, which points to the exit from industrial society. Ulrich Beck (1992) has argued that issues of environmental risk, especially risk related to chemical pollution, toxic wastes, nuclear energy, and biotechnology, call into question the very foundations of industrial society. In industrial society, Beck argues, we happily put issues of economic organization and technological change off-limits to conscious and collective human control. For this reason, Beck believes that industrial society was only 'semi-modern,' in that it only partially fulfilled modernity's promise of rational social development. Beck's emerging 'risk society,' in contrast, puts these issues firmly on the agenda. To Beck, the politics of industrial society was mostly about conflict between social classes, and redistributive issues reflecting this conflict between capitalists and workers. In contrast, the politics of the emerging risk society is organized around the environmental risks which industrial society has generated, but with which it has shown itself incapable of dealing. Unlike industrial society's main hazard of poverty, the rich have no immunity from the hazards of risk society. As Beck (1992: 36) puts it, 'smog is democratic.'

The prospects for strong or reflexive ecological modernization are improved to the extent that environmental affairs can be joined to the risk drama portrayed by Beck. But so long as these affairs are treated in more mundane terms of pollution control and management of material flows, weak or techno-corporatist ecological modernization will prevail. Mol (1996: 317) and Blowers (1997) believe this mundane character is inescapable, and so exclude ecological modernization from any contribution to a reflexive modernity. Langhelle (2000) and Pepper (1999) both see more radical potential in sustainable development discourse, for all its faults. However, sustainable development for Langhelle comes in its Nordic version, true to the spirit of Brundtland rather than the World Business Council for Sustainable Development version. For Pepper, a now-marginalized conception of 'strong sustainability' is the key.

In the weak or techno-corporatist version of ecological modernization, government, corporate capitalism, and the scientific establishment manage the transition to a more environmentally sensitive economic system. But in Beck's risk society, these three institutions share only public disgust for their complicity in the production of risks. Beck sees scientists as risk apologists, their work for sale to the highest bidder. Thus the dominant institutions of industrial society lose their legitimacy in the eyes of the public. Risk society is in fact conducive to the insurgency of a whole new set of interlinked democratic institutions. Experts would lose their privilege, and science would be reformed such that 'research will fundamentally take account of the public's questions and be addressed to them' (Beck, 1999: 70), enabling citizens to reach their own judgments on technical issues. Authority in general would be reconstituted in networks that would cross traditional boundaries of the state, economy, and society. These networks would be the institutions of a reflexive modernity (see also Beck et al., 1994). They could resemble a radicalized version of the governance networks discussed in Chapters 4 and 5.

Ecological modernization in the balance

If ecological modernization does prevail, which kind will it be? Will we get environmentally sensitive management of technological change? Or will we see instead wholesale transformation of the capitalist political economy, the doorway to a reflexive ecological modernity in which the latent human potential for full control of our destiny comes into view for the first time in history? The jury is still out. When it comes to the prospects for this strong or reflexive version of ecological modernization, Beck overstates his case that the transition from industrial society to risk society has occurred. Politics is still mostly about the distribution of material rewards rather than about the production, allocation, amelioration, and distribution of risks, even though the occasional risk issue, such as mad cow disease (BSE) in Britain in the late 1990s or genetically modified organisms, occasionally rises to the top of the political agenda. Moreover, if and when risk society does arrive, it will not necessarily be as conducive to broad-ranging democratization as Beck hopes. For risks can be distributed along class lines, as the environmental justice movement in the United States emphasizes. This movement begins from the recognition that toxic waste dumps and other noxious facilities are normally located in the vicinity of the poor and ethnic minorities. The rich can escape the risks of mad cow disease and genetically modified food by buying organic food. Weak ecological modernization in the wealthy countries could be bought by transferring risks to poor countries—by locating polluting industries in poor countries, or exporting wastes to them, or exploiting their resources in unsustainable fashion. Japan's ecological footprint is very large, but the negative effects are felt mostly outside Japan, in destroyed tropical forests in Southeast Asia, in Pacific islands covered in golf courses, in depleted ocean fisheries, in polluting industries relocated to other countries.

Of the six countries I have emphasized, glimpses of strong ecological modernization can be seen mainly in Germany, where consensual elite-level politics confronts strong oppositional social movements. Ecological research institutes such as the Institute for Applied Ecology (and more than eighty others) supply Beck's 'counter-expertise' for the public, and raise structural questions in their influence upon public policy. The Green Party in the federal governing coalition has negotiated a planned phase-out of nuclear power. German environmentalists have become major players without having to suppress radicalism (the increasing moderation of the Green Party notwithstanding), which has been the price of inclusion elsewhere (Dryzek et al., 2003: 185–91).

I argued at the end of the previous chapter that sustainable development fits uneasily in a world seemingly committed to free trade and the deregulation of markets (unless the discourse is bent heavily in the direction of market liberalism). Ecological modernization faces even greater problems here, given its commitment to conscious collective control of the political economy in the ecological restructuring of capitalism. However, states that operate along these lines might find that they can obtain a competitive edge in the emerging world economic order, if there is money to be made in environmental conservation.

Concerted pursuit of ecological modernization requires a consensual and interventionist policy style consistent with corporatism. This style is, however, anathema to governments under the sway of market liberal doctrines, which helps explain why ecological modernization faces an uphill struggle in the English-speaking industrialized nations. González (2002) points out that at least one element of ecological modernization has been present in the United States since the late nineteenth century, as local economic elites have sought to control pollution for the sake of local economic advantage. Similarly, Scheinberg (2003) suggests looking at the local level, for example at recycling. More far-reaching ideas appeared in former Vice-President Al Gore's book, *Earth in the Balance* (1992), whose proposals are essentially consistent with ecological modernization (though he never uses the term). Gore the Vice-President proved a pale shadow of Gore the author, and none of the book's prescriptions found any place in policy making, still less his 2000 presidential campaign. The disillusioned environmentalists who greeted his public appearances with a chant of 'Read your book!' had a point. Targeting US industry rather than government, Hawken et al. (1999) in *Natural Capitalism* argue that technological measures can enable reductions in resource use and environmental stress while economic growth continues. These US developments remain fragmentary and limited.

The United Kingdom for its part did eventually begin to consider some aspects of ecological modernization in the late 1990s. Fabian Society Director Michael Jacobs (1999) sketched the idea of 'environmental modernisation,' and tried to link it to the general modernization agenda of Tony Blair's Labour government. In 2000 Blair responded, speaking of 'a new coalition for the environment . . . that harnesses consumer demand for a better environment, and encourages business to see the profit of new technologies.' As Barry (2003: 199) points out, Blair's 'new coalition' apparently does not contain environmentalists. In 1998 the UK's Royal Commission on Environmental Pollution published a report, *Setting Environmental Standards*, which remarkably (for a British report) recommended abandoning the established secretive and informal approach to regulation in favor of a more participatory process involving citizens with conscious scrutiny of public values (Weale, 2001: 362–8). This report is actually consistent with strong ecological modernization, but the response from government was tepid, and its recommendations were not implemented.

Such straws in the wind notwithstanding, if the ecological modernizers are right, the United States, United Kingdom, along with Canada, Australia, and New Zealand, are going to be left standing in the transition to a new green capitalist era.

The idea that capitalism's future might be green and corporatist is ridiculed by Prometheans. They regard ecological modernization's precautionary principle as tantamount to lunacy (see, for example, Wildavsky, 1995: 427–33), as guaranteeing only that the wealth which is the real key to environmental health will be dissipated by excessive and costly regulations. Economic rationalists see in corporatism only opportunities for special interests to conspire against the public good. However, policy instruments developed by economic rationalists, notably quasi-market incentive mechanisms (as discussed in Chapter 6), have often found favor among ecological modernizers, who paradoxically find it much easier to implement them because they can strip the instruments of their 'free market environmentalism' ideological baggage.

Green radicals are uneasy with ecological modernization because it threatens to deflect their critiques of industrial society. Ecological modernization might pave the way for the inclusion of green groups in policy making, but at the price of their moderation (Barry, 2003: 204–6). The strong reflexive version of ecological modernization could be stretched to encompass green radical views, though some more romantic green notions would have to be jettisoned to fit this very rationalistic discourse. However far it is stretched, ecological modernization does not easily admit the idea that nature might have intrinsic value beyond its material uses, or green desires for living simply upon the earth in convivial fashion. Human life on earth for ecological modernizers is always going to be a complicated affair, and will never be for living simply.

Survivalists, Prometheans, economic rationalists, and green romantics are probably never going to accommodate themselves to ecological modernization. Governments that have always resisted consensual and corporatist policy making will probably also continue to resist ecological modernization. The discourse at the moment has little to offer Third World societies in terms of developmental alternatives. And it is therefore largely silent on what to do at the global level.

Still, in its weak and techno-corporatist senses, ecological modernization has already proved itself in the cleanest and greenest developed nations. Strong ecological modernization linked to a reflexive modernity is both more intriguing and more speculative. Alone among the discourses surveyed here, it offers a plausible strategy for transforming industrial society into a radically different and more environmentally defensible (but still capitalist) alternative.

NOTES

1 Available at **www.ciesin.org.indicators/ESI/rank.html**.

2 For an earlier comparative study, see Jänicke, 1992.

3 'Green Dreams: Japan,' *The Economist* 362 (12 January 2002): 40–1.

4 Survivalists might remain unconvinced by even demonstrably successful ecological modernization. For even if the rate of increase of resource depletion slows to zero, the depletion is still occurring, and so eventually the resources will run out.

5 Ecological modernization can also be treated as a social science theory for the analysis of environmental developments (Mol and Spaargaren, 2000). As a theory, it is often over-stretched, interpreting any kind of environmental protection as a step on the ecological modernization road; this is especially true when it is applied to societies in the Third World and post-communist world. In this chapter I confine the discussion to ecological modernization as a discourse, not a theory.

6 Hajer (1995: 65) defines the 'credible and attractive story-lines' of ecological modernization as 'the regulation of the environmental problem appears as a positive-sum game; pollution is a matter of inefficiency; nature has a balance that should be respected; anticipation is better than cure; and sustainable development is the alternative to the previous path of defiling growth.'

PART V

GREEN RADICALISM

As befits its imaginative and radical leanings, the world of green discourse is a diverse and lively place, home to a wide variety of ideologies, parties, movements, groups, and thinkers. Found here are green parties and their factions, animal liberationists, bioregionalists, ecofeminists, deep ecologists, social ecologists, eco-Marxists, eco-socialists, eco-anarchists, ecological Christians, Buddhists, Taoists, pagans, environmental justice advocates, green economists, critical theorists, postmodernists, and many others. This variety makes classification difficult. However, green radicalism can be divided into two categories: one that focuses on changed consciousness, another that looks more explicitly to green politics. A stress on green consciousness means that the way people experience and regard the world in which they live, and each other, is the key to green change. Once consciousness has changed in an appropriate direction, then policies, social structures, institutions, and economic systems are expected to fall into place. This prioritization of consciousness is widespread in the green movement, among deep ecologists, bioregionalists, ecofeminists, eco-theologists, and lifestyle greens, among others. Other greens are more attuned to the need to target recalcitrant social, economic, and political structures and practices more directly. They include green parties, social ecologists, eco-socialists, and environmental justice, Third World, and anti-globalization activists. Sometimes the difference between green consciousness and green politics is just a matter of emphasis, and the two join to constitute a green public sphere. Some greens endeavor to combine consciousness change and political change. At other times some contrasts come into play.

These two aspects of green radicalism represent the main options for any social movement. Movements aim to change both the way people think and so behave on the one hand, and social institutions and collective decisions on the other. These institutions and decisions include governments and their policies, though social movements can also target international organizations and corporations, and even help create alternatives to the formal structures of government.

9

Changing People: Green Consciousness

One route to changing the world is through the way people think. In an environmental context, this change would involve the way people experience the world, and the inculcation of new kinds of ecological sensibility. The precise content of this sensibility is contested. In some cases it is radically new, looking to innovative notions of ecological citizenship. In other cases it is radically old, looking back to primal human society before the rise of agriculture. One Earth First! slogan declares 'Back to the Pleistocene!' Sometimes, the radically new and radically old are combined in creative fashion.

The varieties of green consciousness

Deep ecology

Deep ecology as a movement and a label is most prevalent in the United States, though its origins are Norwegian, and it has adherents in Canada, Australia, New Zealand, and elsewhere. Deep ecology was given its name and its initial content by the Norwegian philosopher Arne Naess (1973), who drew a contrast with the 'shallow' ecology movement that only wanted to reform some of the practices of industrial society (see also Naess, 1989). Subsequent development occurred largely in the United States, especially the Western states (see especially Devall and Sessions, 1985), where it became associated with the radical wilderness defense group Earth First!, and with nature writers such as Edward Abbey and Barry Lopez.

According to Devall and Sessions (1985: 67), deep ecology's two basic principles are self-realization and biocentric equality. Self-realization means identification with a larger organic 'Self' beyond the individual person; or 'self-in-Self,' as they put it. The idea is to cultivate a deep consciousness and awareness of organic unity, of the holistic nature of the ecological webs in which every individual is enmeshed. Along these lines, Warwick Fox (1990) describes a 'transpersonal ecology,' a psychological condition of identification and care for other beings, ecosystems, and nature in its entirety. Deep ecologists value species, populations, and ecosystems, not just individual creatures. Biocentric equality means that no species, including the human species, is regarded as more valuable or in any sense higher than any other species. The effective opposite of biocentric equality is anthropocentric arrogance.

Devall and Sessions (1985: 70) elaborate on these two basic principles to argue that nature and its diversity have intrinsic value irrespective of human uses and interests. Given current excessive ecological stress imposed by humans, respect for nature and its diversity requires a reduction in human populations. Unlike some other deep ecologists, Devall and Sessions are not misanthropic, and indeed allow that nature's diversity can be legitimately depleted in order to satisfy 'vital' human needs. It is probably fair to say that their conception of vital human needs does not extend to sports utility vehicles, speedboats, vacation homes, or home entertainment systems.

More misanthropic deep ecologists, most notoriously the pseudonymous columnist in the Earth First! Journal, Miss Ann Thropy, deny the legitimacy of special human interests.[1] In a 1987 article Miss Ann Thropy welcomed famine and disease (such as AIDS) as useful checks on human numbers. The splinter group VHEMT (pronounced vehement), the Voluntary Human Extinction Movement, sounds misanthropic, but its slogan, 'May we live long and die out,' implies only that people voluntarily stop breeding so that the biosphere can recover from us.[2]

Many deep ecologists take pains to distance themselves from the misanthropes. All seek a major reduction in human arrogance when it comes to dealing with the natural world. Most would probably agree with Eckersley (1992: 46), who in defining ecocentrism (roughly synonymous with biocentrism) specifies that it 'recognises the full range of human interests in the nonhuman world' as well as 'the interests of the nonhuman community.'

The question of how to balance human and nonhuman interests is perhaps more easily answered in particular cases rather than at the level of philosophical abstraction. Philosophical dispute about the relative worth of human beings and the smallpox virus does not get in the way of the recognized need to protect the remnant ancient forests of California, Oregon, Washington, and British Columbia against logging; to keep uranium mines out of national parks; and to return the Colorado River to its free-flowing state.

Deep ecologists are quite clear on what to do when it comes to wilderness: preserve, expand, and protect it. 'Big wilderness areas are not only necessary for inspiration and a true wilderness experience, but they are absolutely necessary for the preservation and restoration of ecological integrity, native species diversity, and evolution' (Foreman, 2000: 38). Against social constructionist approaches to nature (see Chapter 1), they are adamant that wilderness has a real existence that predates human appropriation, and represents the only authentic essence of nature (Foreman, 1998). Deep ecologists have much less to say on other environmental issues, such as air and water pollution in urban areas. Urban agglomerations are by definition outside the bounds of defensible human–nature interactions, and thus of no concern. In Edward Abbey's wonderful deep ecological novel *The Monkey Wrench Gang* (1975), one of the heroes, Hayduke, measures road distances in terms of six-packs of beer the driver needs to consume. Hayduke throws the empty cans out of the window: if the environment has already been trashed by a road, a few beer cans make no essential difference.

Ecofeminism

Ecofeminism is a deep philosophy in the sense that it seeks radical changes in ecological consciousness, though it is generally quite hostile to deep ecology. Ecofeminists attack deep ecologists for consorting with macho mountain men such as the fictional Hayduke and the real-world Dave Foreman, co-founder of Earth First! In this light, deep ecology is a doctrine for redneck male adventurers. Worse, its basic diagnosis is wrong. The root of all environmental problems, according to ecofeminists, is not anthropocentrism (human domination of nature), but rather androcentrism (male domination of everything). According to ecofeminists, things begin to go drastically wrong in the way humans treat each other and the natural world with the rise of patriarchy, which dominates women and nature alike. Thus the liberation of women is tied up with liberation of nature; both depend upon the abolition of patriarchy (see Diamond and Orenstein, 1990; Plant, 1989). Patriarchy is seen as cultural rather than natural, and ecofeminists look back to egalitarian and matriarchal societies, some complete with goddesses, prior to the rise of cities, kingdoms, and empires.

Ecofeminism differs further from deep ecology in its sympathy with animal liberation (for example, Kheel, 1990). Deep ecologists, in contrast, show no concern for animals once they are out of nature—for example, in factory farms or laboratories. Even animals in nature are seen as (literally) fair game for deep ecologists, to be hunted and eaten as an expression of the proper human place in ecosystems. Only organic wholes such as ecosystems are to be preserved, not individual creatures. Another point of divergence is on population control. Deep ecologists see a reduction in human population as essential. Ecofeminists believe such a reduction is likely to be accomplished only by further repression and control of women's fertility by the male power structure and its technology (Diamond, 1994).

For all their differences, deep ecologists and ecofeminists alike believe in cultivation of radically different human sensibilities, involving a non-instrumental and non-dominating, more empathetic and intuitive relationship to nature. Both camps are also home to those who advocate a nature-based spirituality, with divinity located in this world, rather than in (male) figures located off the planet (for deep ecology, see Devall and Sessions, 1985: 8, 90–1, 100–1; and Fox, 1984: 203–4; but note that Foreman, 1991: 46 is an atheist. For ecofeminism, see Christ, 1990). Spiritual ecofeminists often look to pagan religions, and are attracted to goddess imagery (for ecofeminist witchcraft, see Starhawk, 1987).

Ecofeminism began in France in 1972 with the formation of Ecologie-Féminisme by Françoise d'Eaubonne, and flourished in the United States in the 1980s. Its Third World dimension is well represented by arguably the world's most prominent ecofeminist, Vandana Shiva, who writes from India. The root of most contemporary social and ecological evils is, according to Shiva, the Enlightenment commitment to science and economic growth, which together destroy life's diversity and sanctity (Shiva, 2000). Shiva is especially concerned by the degree to which imported agricultural and industrial technologies driven by globalization (most recently, genetically modified organisms) further disadvantage Third World women (for example, by denigrating their traditional knowledge of the land and its workings). She points to the leading role played by women in Third World environmental movements, notably the Chipko tree-protection movement in India.

Ecofeminists believe that the kind of sensibility they advocate is not going to be easy for men to adopt. Women are often seen as closer to nature by the fact of their ability to give birth and nurture children (though see Biehl, 1991 for a critique of this dualism). It is the female virtues related to care, empathy, intuition, connection, and cooperation that are crucial. As Plumwood (1993: 9) puts it, many ecofeminists cultivate the myth of a female 'angel in the ecosystem' (Plumwood herself renounces this myth). Men might be able to reason their way toward ecofeminism, but they could never *feel* it. And it is a one-sided emphasis on reason that helped get humanity into such ecological trouble in the first place, by making men so arrogant as to think they could use reason to control nature.

Most ecofeminism is 'cultural' in that it begins with condemnation of one kind of human sensibility and ends with advocacy of another. A more 'social' ecofeminism attends to how society is organized (Carlassare, 1994). The work of Shiva combines cultural and social aspects. Another good example of this kind of analysis can be found in Plumwood's (1995) scrutiny of alternative models of democracy. Plumwood argues that so long as it remains under the sway of liberalism, democracy can never extend itself in a truly ecological direction. For inherent in liberalism are assumptions about the degree to which individuals are properly isolated from one another, rational in a narrowly instrumental and egoistic sense, and unequal in both material wealth and the capacity to exercise power and reason. The consequences are both socially unjust and environmentally destructive. Plumwood's proposed ecofeminist democracy would involve social and ecological citizenship more attuned to an ethic of care and responsibility, together with a more egalitarian political order extending to equality across humanity and nature. However, cultural change still matters for Plumwood (2002).

Bioregionalism

Bioregionalists seek the reinhabitation of places in which people live (McGinnis, 1998). Bioregions can be defined in different ways: by watershed, or by predominant vegetation type. Examples of bioregions would be Pacific Cascadia, covering the coastal forests west of the Cascade Crest from southern British Columbia to northern California; and the Murray–Darling River Basin in Australia. As a movement, bioregionalism is predominantly North American (Sale, 1985). Central to bioregionalism is cultivation of a sense of place. People who live in a bioregion need to adopt it as their true home, to be respected and sustained so that the region in turn can sustain human health and life (Dodge, 1981). They need to become aware of the kind of ecosystem they inhabit, and regard themselves as a part of it, rather than identify with ethnic groups or nations or other human groupings that transcend ecological boundaries. Bioregional consciousness is, then, very different from the kind of consciousness inculcated by the capitalist economy and the mass media, which are destroying regional identity of any kind. Both deep ecological consciousness and ecofeminist consciousness could fit quite easily into a bioregional setting.

Ultimately bioregionalists want to replace local, state, and national political institutions with governments organized along bioregional lines, foreshadowed by existing river basin authorities, which normally, however, lack the right consciousness. Bioregionalists point to the inability of bureaucracies organized on non-bioregional bases to cope with complex ecosystems. A standard example concerns the number of jurisdictions through which salmon in the Columbia River basin must swim between the ocean and their spawning grounds. Proposals for reorganized governments include the 'Europe of the regions' advocated by European greens, which would dissolve existing national boundaries. In North America, Pacific Cascadia might govern what is now Western British Columbia, Oregon, and Washington, and a portion of northwestern California. Ecological concerns would be by definition at the forefront of the agenda of such governments, addressed across all policy areas.

Societies whose focus is provided by the ecosystems which they inhabit, and upon which they must rely for their sustenance, would have to care for those ecosystems very carefully. This assumes that trade across regions would be regulated, presumably by agreement across bioregions. The basic questions of how exactly to draw boundaries and how large bioregions should be remain unresolved. In terms of size, bioregions may be nested within one another. In terms of where to draw boundaries, vegetation type, terrain, human culture, and watersheds may give different answers, and compromises may be necessary across these principles. For example, Pacific Cascadia is a bioregion defined by temperate Douglas Fir forests. But parts of this bioregion fall within the Columbia River Basin—which also contains mid-continent deserts. Many bioregionalists would say such issues are secondary, and that solutions would fall into place once the necessary consciousness were disseminated widely enough.

Ecological citizenship

Bioregional consciousness implies a kind of ecological citizenship, in which individuals learn to become respectful citizens of an ecological place, rather than transforming the place to suit themselves. Such citizenship involves awareness of how the ecosystem supports life, and of life's vulnerabilities. It involves meeting one's material as well as spiritual needs from the resources available locally. However, ecological citizenship does not have to be linked to bioregionalism and obligations to place. For Christoff (1996*b*), such citizenship is equally a matter of commitment to a stewardship ethic and obligations to future generations and other species, irrespective of where they live. Dobson (2004) stresses the obligations of the wealthy, who have already imposed excessive demands upon 'ecological space,' to the poor who have little ecological space. Such space extends to wherever the impact of one's consumption is felt. Ethics of compassion for other humans help, but really such obligation ought to be a matter of justice, not charity. Against economic rationalists, Dobson insists that a sustainable society can be built only by ecologically motivated citizens, and not by consumers and producers responding to the carrots and sticks of economic incentives.

Lifestyle greens

For some in the movement, the essence of being green is not adherence to any philosophical analysis of the sort favored by deep ecologists and ecofeminists, still less any kind of collective action, political or otherwise.

Instead, being green is a matter of leading a green lifestyle. To some, it is mostly a matter of green consumerism: buying Body Shop or Aveda cosmetics whose constituent chemicals are not tested on animals, eating vegetables grown organically, boycotting genetically modified organisms, buying biodegradable cleaning products, using toilet paper made from recycled fibers, and resisting the blandishments of corporate advertisers (especially when they attempt to 'greenwash' their products). The green lifestyle is often vegetarian, for the vegetarian's caloric intake imposes much less stress on agricultural land than does flesh-eating. Recycling, composting, and bicycling as opposed to driving a car all play their parts.

Such decisions can be instrumentally good for the environment, and may even make good economic sense. So local councils throughout the world have adopted recycling for both environmental and economic reasons. Yet lifestyle greens are concerned with much more than the immediate piecemeal effects of such decisions. It is not just a matter of doing green things, it is a matter of being green in doing them, of using these actions to cultivate a post-industrial way of experiencing and relating to the world.

Eco-theology

I have already noted that deep ecologists and ecofeminists alike are often attracted by some kind of nature-based spirituality (see also La Chapelle, 1978; Spretnack, 1986). However, these two movements do not exhaust the range of eco-theology (for comprehensive catalogues, see Gottleib, 1996; Hay, 2002: 94–119; and, for updates, the journal *Ecotheology*). Eco-theologians diagnose the root of environmental problems in spiritual terms, and if the root of the problem is spiritual, then so too must be the cure. The classic argument here is that of the historian Lynn White (1967), who argues that environmental crisis is the product of Judeo-Christian religious tradition, which places god outside of and above nature, and then proclaims that man is made in god's image. This placement provides justification for unlimited human manipulation and abuse of nature for purely human ends. White does see in the Judeo-Christian tradition an alternative possibility, associated most prominently with St Francis of Assisi, who White proposes as the patron saint of ecology.

Other eco-theologists are more inclined to give up on Judaism and Christianity in favor of Eastern religions such as Taoism, Buddhism, and Hinduism, all of which cast humanity in far more humble terms, adopting a contemplative and reverential attitude toward nature. E. F. Schumacher's (1973) 'Buddhist economics' would be built on individuals who seek to maximize wellbeing at a minimum level of consumption. Critics point out that the societies in which these religions flourish are no less prone to ecological destruction than Western societies. Moreover, they also seem to go hand-in-hand with mass political passivity, which might severely hinder any attempt to reorient society in a more environmentally sound direction through grassroots action.

Eco-theology can also enter the analyses and prescriptions of movements and thinkers interested primarily in political change. In 1991, the First National People of Color Environmental Leadership Summit in the United States declared that 'environmental justice affirms the sacredness of Mother Earth.' Two of the prominent survivalists discussed in Chapter 2, Robert Heilbroner and William Ophuls, both argue that the wholesale political transformations they advocate must be accompanied by some kind of spiritual transformation. So Heilbroner (1991: 176–7) speaks of governments of the future that will be 'monastic' in their form, combining 'religious orientation with a military discipline.' Ophuls (1977: 243) for his part believes in the necessity of 'metanoia . . . tantamount to religious conversion.' For Heilbroner, religion's main value is instrumental rather than intrinsic: it is just a way of keeping people in line, to stop them abusing the environment. For Ophuls, it is helpful in making the transition to a different kind of political economy more palatable to the population at large. Along these lines, the leading British Green Jonathan Porritt argues that movement to a sustainable society is unlikely 'without some huge groundswell of spiritual concern' (Porritt, 1986: 210).

The romantic disposition and its critics

A romantic disposition infuses some advocates of green consciousness change. For romantics, politics is not about devising strategies to achieve tangible goals; rather, it is an arena in which different kinds of experiences can be sought and developed. Historically, the term romanticism was attached to an intellectual movement in the eighteenth and nineteenth centuries that opposed the rise of modern science and liberal politics and positioned itself against the Enlightenment. Enlightenment is the name for the eighteenth-century movement that renounced religion, myth, and traditional social order in the name of reason. Reason in turn meant liberal politics and human rights of the sort established in the English, American, and French Revolutions of 1689, 1776, and 1789 respectively. Reason also meant that modern science became the route to secure knowledge, which enabled in turn the growth of modern technology and the manipulation of nature on a large scale. The defining feature of modern society is that it embodies the principles of the Enlightenment. The original romantics favored instead an artistic and aesthetic orientation to life and politics. To romantic poets such as Coleridge, Wordsworth, and Shelley, nature and humanity belonged in an organic relationship best understood and developed through feeling and insight. In the United States, the romantic disposition was developed in the nineteenth century by Ralph Waldo Emerson and Henry David Thoreau.

Just like the romantics of the eighteenth and nineteenth centuries, green romanticism rejects core Enlightenment principles, pointing to the environmental destruction caused by modern science and technology wielded in human arrogance (Hay, 2002: 4–11). Green romanticism seeks to save the world by changing the way individuals approach and experience the world, in particular through cultivation of more empathetic and less manipulative orientations toward nature and other people. It is heir to the older romantic rejection of the Enlightenment's emphasis on rationality and progress. This kind of disposition can be found among deep ecologists, though Foreman (2000: 38) rejects 'airy-fairy flights of romantic fantasy.' It is shared by some ecofeminists, especially those who subscribe to what Plumwood criticizes as 'the angel in the ecoystem' image; by those bioregionalists who place intuitive experience of place above all else; and by spiritual greens attuned to oneness with nature.

Other greens caution against abandoning Enlightenment values (Hayward, 1995). Enlightenment does have a dark side, in the form of instrumental reason in the service of anthropocentric arrogance, underwriting uncontrolled economic growth, oblivious to the constraints imposed by the natural world, and to the damage done to conviviality in the social world. The brighter side of Enlightenment involves hostility to unquestioned hierarchy, a corresponding commitment to equality (at least among humans), basic human rights, and the possibility of free dialogue as the essence of rational social relationships. For Enlightenment rationality is not just a matter of manipulating the world on behalf of the mind's desires, as green romantics would have it. Rationality is also a matter of open-ended and critical questioning of values, principles, and ways of life—which opens the door to critical ecological questioning. As Plumwood (1993: 4) puts it: 'critiquing the dominant forms of reason which embody the master identity and oppose themselves to the sphere of nature does not imply abandoning all forms of reason, science, and individuality. Rather, it involves their redefinition or reconstruction in less oppositional and hierarchical ways.' Hay (2002: 10–11) points out that contemporary green radicalism is rooted in science as well as aesthetics, an ecosystemic view rather than romantic individualism, and progress to a better future, not a return to some golden past.

Discourse analysis of green consciousness

The essential storyline of green consciousness change is that industrial society induces a warped conception of persons and their place in the world. Required to remedy this situation are new kinds of human sensibilities, ones that are less destructive to nature. While the precise content of the required sensibilities varies, they would generally involve a less manipulative and more humble human attitude to the natural world and to each other. Digging a little deeper, the storyline is constructed from the following elements.

Basic entities whose existence is recognized or constructed

In the background of green radicalism are global ecological limits of the sort that energize survivalists: the existence and proximity of these limits impart a sense of urgency. As Dobson (1990: 73) puts it: 'the foundation stone of Green politics is the belief that our finite Earth places limits on our industrial growth.' Green radicals would not, however, share the authoritarian political prescriptions favored by some survivalists. Green cultural change would still be coherent and defensible (though less urgent) in the absence of limits.

The basic entity whose existence is recognized and constructed, and which forms the real foundation for the discourse, is nature. Both inner nature (that is, of the mind, body, and spirit) and outer nature are at issue here. Green cultural change would bring these two into closer harmony by operating on human sensibilities and obligations. Such harmony is the essence of deep ecological notions of 'self-in-Self,' and of cultural ecofeminist principles concerning a more intuitive and empathetic human orientation to the natural world—and to other humans. Opposed to these conceptions of nature and the natural are notions of the unnatural. Included here would be the core practices and—still more important—the core sensibilities that industrial society has inculcated in people. Such unnatural acts and orientations would include the anthropocentric arrogance of Prometheans, the relentless quest for more and better consumer goods, and the instrumental calculation emphasized by economic rationalism. Ecofeminists would add patriarchy to this list.

An emphasis on consciousness change means that social structures, institutions, and policies are seen as having no independent life of their own, but instead as ultimately reducible to the underlying sensibilities of the members of society. Thus does green consciousness fall squarely in the philosophical tradition of idealism (as opposed to materialism, which stresses the causal influence of economic forces in shaping society). It is ideas, not material forces, that move history: so the key to changing the world is to change ideas.

Assumptions about natural relationships

The natural relationships stressed by advocates of green consciousness change are often, quite simply, natural relationships. The images of nature and the kinds of relationships it contains may vary in their details. For example, deep ecologists are often quite happy with a nature 'red in tooth and claw,' to use an expression favored by some followers of Charles Darwin, and celebrate predation and hunting as parts of the natural order. Ecofeminists are more likely to see in the natural world harmony between creatures and species. Whatever its balance of competition and cooperation, the natural order is an egalitarian one (the deep ecologists' biocentric equality), in which there is no hierarchy, and certainly not a hierarchy that puts humans on top of everything else. Green radicals also believe that this order has been violated by humankind, be it through anthropocentric arrogance, patriarchy, or industrialist indifference.

Such violation often comes in the name of rationality. The dominant form of rationality in today's world is instrumental rationality: the capacity to devise, select, and effect good means to clarified and consistent ends. Instrumental rationality calls up a dichotomy between subject and object. Only the human mind is subject. Everything else, including the natural world and disadvantaged humans, consists of objects to be manipulated and dominated for the sake of whatever the mind desires. Thus instrumental rationality estranges us from nature and each other, with all kinds of disastrous consequences. Nature takes its revenge for our arrogance by inflicting environmental crises upon us.

Agents and their motives

The cultural aspect of green radicalism is about the cultivation of alternative kinds of ecological subjectivity. Every person can be an agent, with the capacity to craft his or her own relationships to the natural and human world. This discourse is, then, well populated by human subjects. More important, it contains only subjects. People are subject to themselves in that it is up to all of them to create an appropriate orientation to life; nobody else can do it for them. Collective actors such as governments, corporations, and other organizations are largely ignored, except to be condemned, as are elites who might have the power to impose their will on other people. However, the discourse does allow for some human subjects to be more enlightened than others, and so show the way. Thus there is scope for green education (see Dobson, 2004: 174–207)

The ascription of agency does not stop with human beings. In a rejection of the weight of several hundred years of natural science, and of dominant notions of instrumental rationality, agency can be seen as existing in (external) nature too. This is especially true for the romantic disposition in green consciousness change. Nature does not have to be blind, unthinking, and unfeeling; instead, it can be truly alive with meaning and purpose. This applies to individual creatures, to species, to ecosystems, and perhaps even to the planet as a whole. James Lovelock's

(1979) Gaia hypothesis has found a sympathetic reception among greens (though as a persistent enthusiast for nuclear power he has hardly sought such a reception). This hypothesis states that the biosphere in its totality acts collectively to maintain the conditions for life on Earth. In this light, the biosphere is a self-regulating entity that can correct for threats to its capacity to support life—threats that come, for example, through increases in the level of solar radiation, volcanic activity, or human pollution. Acceptance of this hypothesis does not guarantee an environmentalist outlook. For Gaia may be quite able to correct for any abuses that we humans can dream up, be it pollution from burning fossil fuels or nuclear holocaust. We may wipe ourselves out, but Gaia will persist, just as it has outlasted the extinction of millions of species. Green radicals are far more likely to regard Gaia as both vulnerable and worthy of reverence (see, for example, Porritt, 1986: 206–9).

Key metaphors and other rhetorical devices

Green advocates of consciousness change make use of an eclectic range of biological and organic metaphors. Given the focus on the cultivation of human subjectivity, many of these metaphors are incorporated into exhortations about how to experience the world. For example, according to Robert Aitken, 'Deep ecology . . . requires openness to the black bear, becoming truly intimate with the black bear, so that honey dribbles down your fur coat as you catch the bus to work' (quoted in Dobson, 1990: 61). To Karen Davis (1995), the key to a feminist understanding of animal liberation is 'thinking like a chicken.' To use a popular expression in deep ecology, the key is instead 'thinking like a mountain.' Dave Foreman, deep ecological founder of Earth First!, used to conclude his standard stump speech by getting his audience to howl like wolves.

Changing the way people experience the world is crucial. This can be done through argument; and there are indeed many detailed arguments, some of considerable philosophical sophistication, on behalf of the various green positions I have surveyed. But ultimately, argument may not enough. Reasoned argument can only take us so far along the road. The rest of the path may require rhetorical strategies that reach beyond reason to passion. If the point is to convince listeners of the desirability of an intuitive and empathetic orientation to nature, then this can be done by relating personal stories, analogous to accounts of religious conversion and how it changed the life of the teller of the story. So, for example, the ecofeminist Julia Russell (1990: 224) relates how she came to the realization that the Earth is a living being through contemplation of her compost pile, which showed her that 'The Earth turns everything given to it into itself.' Appeals can be and are made to intuitions and emotions. Poetry, art, religious and quasi-religious ceremonies, the telling and re-telling of myths and creation stories can all play their parts. Dobson (2004: 211) suggests that 'an hour's lived experience can produce more politicisation than a year in class' when it comes to inculcating a sense of ecological citizenship.

The impact of green consciousness change

For most of the discourses surveyed in earlier chapters, it makes sense to look for real-world impacts in terms of the policies of governments and international bodies, and in the reconstruction of social, economic, and political institutions and practices. But to assess green consciousness change in similar terms would be to miss its central point. For these greens want *people* to be different; and when they are, then everything else is

BOX 9.1 **Discourse analysis of green consciousness change**

1. **Basic entities recognized or constructed**
   - Global limits
   - Nature
   - Unnatural practices
   - Ideas
2. **Assumptions about natural relationships**
   - Natural relationships between humans and nature that have been violated
   - Equality across people and nature
3. **Agents and their motives**
   - Human subjects, some more ecologically aware than others
   - Agency can exist in nature too
4. **Key metaphors and other rhetorical devices**
   - Wide range of biological and organic metaphors
   - Passion
   - Appeals to emotions, intuitions

expected to fall into place. Many social movements do in fact take effect largely through the changes in culture, ethics, and so people's behavior that the movement induces (Tesh, 1993). Feminism, for example, has successfully altered relationships of power between the sexes within the household, and in society more generally. Some of these changes have been confirmed by legislation such as family law and equal opportunity measures, but many of them have not been so compelled. Similarly, a large part of the impact of the last thirty years of environmentalism is in the way people have come to think about their everyday behavior: in recycling wastes, in insulating their houses, in paying attention to the environmental friendliness of products they purchase, in what they tell their children about the world. Cultural change can also influence the understandings of key decision makers (Wapner, 2002: 53–8), though by the time green ideas get taken up they have often lost much of their radical bite.

Thus it is in this cultural realm that we should seek evidence of the real impact of green ideas. The cultivation of new ways of being is done not just for the benefit of the individual so enlightened, but for the good of society, and ultimately the good of the planet.

In these terms, it is a bit ironic that the main impact so far of green consciousness change is probably at the level of changing consumer behavior, such that, of the varieties of cultural change surveyed earlier, it is the lifestyle greens who have had the most effect. In industrial societies, at least, many people do happily sort and recycle their garbage, read labels of products on the supermarket shelves, shun ozone-depleting chemicals and genetically modified organisms, compost food scraps and garden waste, force companies like McDonald's to stop using Styrofoam packaging, and improve the energy-efficiency of their lifestyles. Here, green cultural change provides useful and perhaps unexpected support for ecological modernization, which requires consumers to behave in exactly this fashion.

What people have not done, except in very small numbers, is adopt the kind of ecological consciousness of the kind sought by deep ecologists, ecofeminists, bioregionalists, or eco-theologians. The relevant groups and networks are often quite small, and not especially visible to a larger public. The most noteworthy such group is probably Earth First!, founded in the United States in 1980. Earth First! is perhaps known less for its deep ecological philosophy as for the exploits of its members. These exploits include lying down in front of bulldozers, gatecrashing the anniversary celebrations of Lake Powell reservoir on the Colorado River, occupying the tops of trees in old growth forests scheduled for clear-cutting, putting a 'crack' on the face of the Glen Canyon Dam, and so forth. Earth First! is also associated with monkeywrenching or ecotage: that is, sabotage of environmentally damaging activities. Monkeywrenching figures large in the rhetoric of Earth First! supporters and opponents alike (for a field guide, see Foreman, 1985). The possibilities include pouring emery powder into the crank-cases of earth-moving machines, pouring syrup into fuel tanks, hammering spikes into trees to make it dangerous to cut them, pulling up survey stakes, and destroying logging roads. Since 1997, the impetus has been taken up by the Earth Liberation Front, whose supporters torched a ski lodge under construction in Vail, Colorado, in 1998, and set fire to environmentally destructive sports utility vehicles in dealerships.

Monkeywrenching in fiction is far more thoroughgoing: the best accounts remain those of the deep ecological nature writer and novelist Edward Abbey (1975, 1990). The rhetoric and fiction have obviously influenced the United States Federal Bureau of Investigation, which devoted a great deal of effort to infiltrating Earth First! and entrapping some of its members, including Dave Foreman, in a 1989 plot to blow up power lines. The FBI eventually classified the Earth Liberation Front as the number one terrorist group in the United States (even though no person has been injured as a result of an ELF action). The FBI has been far less diligent in seeking out the perpetrators of the real violence surrounding deep ecology, which comes almost entirely from the anti-ecological side. For example, when a bomb in her car injured Earth First! activist Judi Bari, the authorities initially described it as a case of the activists being blown up by their own bomb; when that account was discredited, no proper effort was made to catch the real bombers. When in 1998 Earth First!er David Chain was killed by a logger felling a tree on him in California's Headwaters Forest, no charges were laid.

Many of those who share green sensibilities do not belong to any group. Many belong to more conventional environmental interest groups. David Brower was perhaps the most visible, militant, and influential figure in conventional environmentalism in the United States from the 1950s to the 1990s, leading the Sierra Club, founding the League of Conservation Voters,

Friends of the Earth, and the Earth Island Institute. Green romanticism was quite explicit in his speeches and writings. Not for nothing was he known as the Archdruid (McPhee, 1970).

Green consciousness can also be cultivated in what Torgerson (1999) calls the 'carnival' aspect of green politics. Here, industrialism is opposed not just by serious political action, but also by green comedy. Such comedy might include Earth First!-style tree sitting in the United States, Canada, and Australia; the digging and inhabitation of precarious tunnels in the path of motorways by anti-roads activists in Britain; Greenpeace stunts such as installing solar panels on the roof of the house of Australian prime minister John Howard, a die-hard fossil fuel promoter; and Reclaim the Streets events in London that shut down traffic. Torgerson's carnival would inspire creative thinking about the place of people in their world. If nothing else, green comedy would help prevent greens becoming like their opponents: constipated, serious, strategic, calculating, and grey.

Can green consciousness save the Earth?

Green radicals believe the world needs saving. But is consciousness change up to the task? There are several reasons why it may fall short. The first problem is the practical one of convincing large numbers of people to change the way they relate to the world. How exactly is the green vanguard to convince everyone else? The normal answer comes down to the vanguard educating everyone else in how to think and act in ecologically defensible fashion. Surveys show that most people, at least in developed countries, already consider themselves environmentalists. Among people under the age of 30 in the United States, this figure is as high as 85 per cent (Thiele, 1999: 211). But such commitments are often shallow. While widespread in American high schools, environmental education does not challenge entrenched industrialist worldviews, and imparts only a very thin conception of ecology (Bowers, 1999).

A further problem arises in connection with complexity in ecological affairs. The biologist and environmental activist Barry Commoner argued long ago (1972) that the first law of ecology is that 'everything is connected to everything else.' While this is a slight overstatement, there is no denying the inherent complexity of ecological problems (see also

Dryzek, 1987: 28–9). Interventions in complex systems can produce counterintuitive results, however well intentioned the intervention. Thus good intentions and sensibilities are never sufficient as secure guides to action. For example, it was long believed that the best way to protect ecosystems in the forests of the American West was to suppress fires; ecologists later realized that these ecosystems depended for their renewal on periodic burning. Loving the Earth never guarantees that you will treat it well.

Matters here are made more difficult still once environmental affairs are recognized in terms of crisis. Crisis means that human interactions with the natural world are in severe disequilibrium. Now, improved green sensibilities might be sufficient to maintain an equilibrium in which people lived in harmony with nature, but they cannot tell us how to get from our current severe disequilibrium to this harmonious state. There is no theory of the transition, which surely requires some political program, and some kind of action at the collective level. The problem here is that social, political, and economic structure is more than just a reflection of the attitudes of society's masses or elites, and so changed sensibilities will not necessarily lead to structural change.

Why does social structure matter? The main reason is that macro consequences (in terms of policies, institutions, and events such as revolutions) are rarely if ever a simple extrapolation of micro causes. In Buddhist, Taoist, and Hindu societies, pervasive environmentally sensitive sentiments can coexist with despotic and anti-environmental social, political, and economic systems. There are major issues involved in the aggregation of individual-level preferences, attitudes, and sensibilities into macro-level results (see Coleman, 1986). Speaking as a social scientist, I note that social science only exists because societal and social-structural phenomena are not reducible to individual psychology. So even if there were large-scale conversion of individuals along the lines sought by greens, it is quite possible that nothing at all would change at the macro level. If there is no structural setting which facilitates the articulation of frustration with the old order, the construction of solidarity against that order, and action based on that solidarity, then the old order will survive. Mass psychological and cultural changes can have macro-level consequences, but they are never a simple reflection of micro-level transformations. Psychological changes can be frustrated by a host of factors: electoral systems that discriminate against new parties, market systems that reward and reinforce materialistic and egoistic behavior, social structures that isolate individuals and privatize their concerns, employment structures that make it hard to meet and organize, family structures that either keep women in the home or reinforce privatization by making both male and female income-earners too exhausted to have time for political action.

The most important such structural constraint exists in connection with a global liberal capitalist political economy that is more secure and powerful than ever before. This political economy conditions not just structures and institutions, but also identities, subjectivities, and discourses. As Lindblom (1982) notes, the market imprisons government policy: there are certain things governments simply must do as a first priority, notably maintain the confidence of capitalist investors. He adds that the market also imprisons the way most people think: if there is a conflict between market imperatives and other values (including environmental ones), it is generally taken for granted that these other values must give way.

Thus the challenge to greens is: how will your proposed alternative consciousness fare in a world currently structured to guarantee its frustration, and moving in a direction that reinforces such frustration? What aspects of the world are conducive to alternative green subjectivities? What aspects get in the way? What is the relative strength of these enabling and constraining forces? How might political and economic structures be changed so as to alter the balance of these forces? Who or what would resist such changes?

It is exactly such questions that the more explicitly political strand of green radicalism seeks to answer.

NOTES

1 Miss Ann Thropy is actually Christopher Manes, whose own account of radical environmentalism can be found in Manes (1990).

2 See **www.vhemt.org**.

10

Changing Society: Green Politics

Green radicalism is about political change targeted at social structures and institutions as well as consciousness change. This more overtly political emphasis is advanced by a number of movements and schools of thought. Their degree of radicalism varies from eco-anarchists who seek a wholly new and currently very distant kind of society to 'realo' greens who in several countries have achieved a share in the power of government.

The varieties of green politics

Green parties

Green radicalism finds its most conventional form of organization in political parties. Green parties have been part of the electoral landscape for over a quarter of a century, and have in several countries (including Belgium, Finland, France, Germany, and Italy) joined governing coalitions and provided government ministers (especially environment ministers). The German Greens, *Die Grünen*, occupy a central position. The German Green Party was not the world's first—the claimants for that title include the United Tasmania Group in Australia and New Zealand's Values Party, both formed in 1972. But the German Greens have long been regarded as the world's most significant, for reasons relating to their size and success as both a movement and a party. *Die Grünen* were founded in 1980 and entered the federal *Bundestag* in 1983 with 5.6 per cent of the national vote. Their electoral high point came with 8.6 per cent of the vote in 2002, by which time they had been in a governing coalition at the federal level with the Social Democratic Party for four years.

The deep national history with which the German Greens have had to cope is one in which romanticism looms large. The romantic reaction against modernity in the eighteenth and nineteenth centuries was stronger in Germany than elsewhere, and bound up with reactionary German nationalism. Later, this combination would gain an environmental edge in a strand within the Nazi Party, which idealized the Nordic natural environment (the Rhine, Black Forest, Alps, etc.) as well as the Nordic race, proposing a mystical connection between race and environment. As Anna Bramwell (1989) reminds us, the history of European green politics in the first half of the twentieth century is located on the fascist right rather than the progressive left. Mindful of this history, the German Greens are wary of romanticism, and especially suspicious of green spirituality (see Capra and Spretnack, 1984: 53–6).

The German Green Party was long divided into two main factions, the *Realos* and the *Fundis* (the Greens' most well-known figure from the early days, Petra Kelly, was eventually reviled by both factions as soon as she became a media star). *Realos* believe in action through the system, especially through parliamentary politics, accepting the need for a 'long march through the institutions' if green aims are to be furthered (Wiesenthal, 1993). They attend closely to vote-maximizing strategies, party organization, and parliamentary tactics, and are open to coalition with other progressive forces, notably the Social Democratic Party. Such coalitions began in government at the city and *Land* (state) levels. *Fundis*, in contrast, believe that the Greens are properly a social movement rather than a political party, and that it is the green task to confront an irrational political system rather than work within it. The most well-known *Fundi*, Rudolf Bahro, left the party noisily in 1985 in protest against its refusal to issue a blanket condemnation of animal experimentation. The bitter and heated debate between the two factions was largely resolved with the victory of the *Realos* in the early 1990s, under the leadership of Joschka Fischer. But even in his ascendancy Fischer admitted publicly that his main problem was that most members of his party thought he was a jerk. This assessment was not shared by the German electorate. Fischer proved to be a popular foreign minister after the Greens joined a federal governing coalition in 1998, and is widely credited with saving the coalition at the 2002 federal election. By this time Fischer was respected internationally as a leading statesman, a far cry from the fiery radicalism of his youth.

Green *Realos* want to change the world through influence on public policy, not just individual consciousness. Claus Offe's (1990) analysis of the German Greens, which resonates with Green self-conceptions, treats them as a paradigm case of a new social movement. Offe believes that such movements take on historical significance as the third major wave of protest in modern societies. The first wave was liberal capitalist protest against the rigidities of a feudal society governed by aristocracy and monarchy. The second wave involved socialist protest against the victorious liberal capitalist system. And the third is that of the new social movements, encompassing not just greens but also feminists, peace activists, and various urban protests. Such movements are not interested in any romantic return to a pre-modern past. Instead, they are committed to what Offe (1985: 853) calls a 'selective radicalization of modern values,' notably freedom, equality, and democracy. Thus they can be located squarely within the Enlightenment emphasis on social progress. Obviously the Greens regard much of what has transpired in the name of modernity—the destruction of nature, the depletion of resources, the bureaucratization of social relationships—as undesirable. But the solution is not to be found merely in changed individual sensibilities, in any return to a preindustrial Eden, or in postmodern playfulness. Rather, it is to be found in hardheaded analysis of social, political, and economic practice and structure. This analysis in turn can best be developed and put into effect through discursive and democratic interaction within the movement, be it the formalized party organization of the *Realos* or the more fluid and informal groupings favored by the *Fundis*.

The United States Green Party looks very different from its European counterparts, more a loose network than a party, and just as interested in philosophy and lifestyles as in politics. However, in 2000 the Green ticket of Ralph Nader and Winona Laduke won 2.6 per cent of the vote in the presidential election, over 5 per cent in eleven states. Nader was accused of denying victory to Democrat Al Gore, and enabling George W. Bush to be installed under dubious circumstances following electoral irregularities in Florida. The US electoral system discriminates against third parties, which find it almost impossible to secure representation under the single-member constituency, simple-plurality voting rule. Most environmental groups in the United States therefore shun the Green Party, though they are active in electoral politics for or against candidates from the two major parties.

Social ecology

Social ecology is deep ecology's main eco-philosophical rival in United States green circles. It is associated with the veteran eco-anarchist Murray Bookchin, and explicated in fine detail in his numerous writings (for example, Bookchin, 1982, 1990). As its name implies, social ecology emphasizes the 'social' dimension often missing in deep ecology. To Bookchin, the root of all evil, in human society no less than in human relationships with nature, is hierarchy. Hierarchy has arisen only in the last six thousand years or so of human civilization. Whether manifested in the domination of peasants by lords, of women by men, of the countryside by the city, of the young by the old, of workers by capitalists, of society by the state, of nature by people, or of the body by the mind, hierarchy is a profoundly undesirable and unnatural phenomenon. For Bookchin sees no hierarchy in the nonhuman world. Relationships that humans perceive as competitive or dominating are in fact subtle examples of mutual benefit. For example, herbivores benefit from predation by carnivores because it keeps their populations in check, eliminating frail and diseased members. Nature is not the violent struggle for survival of the fittest which apologists for war and capitalism portray. Instead, nature properly understood is a cooperative place, indeed a model for harmonious human society, the place where freedom originates.

At first glance this might look like a recipe for romantic return to a primal Eden. What saves social ecology from this fate is Bookchin's stipulation of a special place for humanity in the natural scheme of things. Humans are not set above nature, as they are in the crude anthropocentrism of Prometheans and economic rationalists. However, they are treated as the only bit of nature that has yet achieved self-consciousness: we are nature become aware of itself. We should not deny this aspect of *our* nature in the name of biocentric egalitarianism. Human social evolution now occurs in what Bookchin calls 'second nature,' an environment that is cultural rather than biological. Thus we should embrace the idea that there is such a thing as progress in human sensibilities. In this light, Bookchin (1986: 75) argues that 'we cannot avoid the use of conventional reason, present-day modes of science, and modern technology.' Certainly modernity has meant that the longstanding human propensity to hierarchy has been able to take new and more insidious forms. The state has been able to perfect its domination over society with the development of bureaucratic rationality, and capitalists have been able to deploy science and technology to dominate workers and further subdue nature. But modernity also brings increased ability for humans to question hierarchy, and to contemplate more open and egalitarian interchanges with each other—and with the nonhuman world. Thus Bookchin (1995) does not hesitate to criticize postmodernists as well as anti-modernists.

Social ecology's place in the modern world is further secured by Bookchin's socialist credentials: he is a veteran leftist as well as a veteran ecologist. Thus social ecology concerns itself with the analysis of the institutions and practices which perpetuate injustice—notably, hierarchy and competition associated with modern state structures and capitalism. Bookchin's anarchist solution of small-scale, mostly self-sufficient local communities existing in harmony with their neighbors and with their local environment may be utopian; but it does at least rest on a political-economic analysis, and proposes a political-economic strategy. Bookchin in his later work and his followers in New England have developed ideas about 'radical municipalism,' which involves the renewal of political institutions from the ground up, starting at the local level.

One other side of social ecology which owes much to Bookchin's background on the radical left is sectarianism. Bookchin reserves his most bitter invective not for oil companies, chemical corporations, or their servants in government, but for other environmentalists. His most frequent target in the 1980s was deep ecology, which he denounced as an ugly wart on the face of the ecology movement, propounded by reactionaries guilty of racism and eco-brutalism. For deep ecology, Edward Abbey responded in kind: the late 1980s saw these two sweet old men laying into one another. Abbey at one point threatened to take a quirt (horsewhip) to Bookchin if he ever showed up in Arizona; though Abbey later relented, saying that a fat old woman like Bookchin had nothing to fear from him. Come the 1990s, reconciliation was in the air: Abbey was dead, and Bookchin and Foreman (1991) showed in amicable debate just how much common ground could be found between deep and social ecology.

It is no accident that in an American context Bookchin's ideals are most at home in the pastoral landscape of Vermont, where an image of humanity and nature in productive harmony is readily envisaged, but where there is no wilderness. In the American West, in contrast, the clash between humanity and nature appears violent and intractable. Some of the wilderness remains, but human economic activity takes the form of clear-cut forests, ravaged grazing land, mining scars upon the desert that do not heal, huge dams that destroy riverine ecosystems. Deep ecology has both its most fervent supporters and its most bitter opponents in the West, which is also home to hard-line anti-environmentalism in the John Wayne tradition, for which nature exists only as a challenge to be conquered.

Red and green

One of the German Green slogans is 'Neither left nor right, but in front.' Not all greens agree, least of all in Germany, where there used to be a sizeable Marxist faction within the Greens. Of course, there is much more to the left than Marxism, and so much more to left green thinking than eco-Marxism.

In the early years of environmental resurgence in the 1970s, Marxists typically denounced environmentalism as bourgeois and concerned only with life's pleasures, a distraction from the real stuff of class struggle. Marx himself was a Promethean, who cared about nature mostly for the sake of its conquest, the best efforts of some contemporary Marxists to rehabilitate his environmental reputation notwithstanding. Matters have now changed, and many Marxists are now eco-Marxists. With capitalism's own dynamics having failed to culminate in socialist revolution as Marxists once predicted, they can now look to ecological crisis as a harbinger of a general crisis of capitalism. On this account, capitalism destroys the ecological base upon which all human economic activity rests, as well as creating a class of workers and others poisoned by pollution. Eco-Marxism departs from other varieties of green radicalism in believing that this destructiveness is contingent upon capitalism, such that a more rational economic system would not be subject to ecological limits. Kovel and Lowy (2002: 156) speak of a 'transformation of needs' that would make limits irrelevant because the economic system would be freed from pursuit of ever-greater quantities of material wealth.

Eco-Marxists see ecological issues as laying bare many of the contradictions of capitalism, and ultimately contributing to its demise, though they are a bit coy on exactly how and when this will happen. Kovel and Lowy (2002: 1) envision an 'ecosocialist international' that would unify many existing local and national struggles that currently do not recognize their links to one another, and to the Marxist critique of capitalism. Any alternative political-economic system to replace liberal capitalism would of course have to avoid the gross environmental failings of the countries of the former Soviet bloc—failings which provide plenty of ammunition to anti-Marxists (for example, Lewis, 1992: 163–6).

Eco-Marxists devote their energies not just to the critique of capitalism, but also to criticism of green thinking that involves blanket condemnation of anthropocentrism or modernity. Eco-Marxists believe that the real explanation for ecological crisis revolves around material economic factors. Human consciousness is relevant only to the extent it can be tied to these forces. Thus they scorn the ecocentric proposals of deep ecologists (Pepper, 1993: 221–5). Eco-Marxism's leading lights include the US sociologist James O'Connor (1988), founder of the red-green journal *Capitalism, Nature, Socialism*,[1] and Joel Kovel (2002).

Eco-socialists who are not eco-Marxists are often proponents of the need for government planning to cure the ecological irrationalities of capitalism (see, for example, Ryle, 1988; Stretton, 1976). There is no suggestion that capitalism needs to be superseded, merely that it needs to be tamed so that it uses resources less wastefully and involves less wanton environmental destruction. The distinction between eco-socialism and administrative rationalism as discussed in Chapter 4 is sometimes hard to sustain. Certainly, eco-socialists would happily use the same range of policy instruments as administrative rationalists, though their goals would include a broader range of social justice concerns. This emphasis on state planning might distress green *Fundis* whose version of green politics would oppose the state rather than use it. Social ecologists, bioregionalists, and other greens believing in economic and political decentralization would likewise be skeptical about eco-socialism. Yet neither centralization nor decentralization is a litmus test for green radicalism. Even eco-anarchists recognize the need for some kind of authority above the local level, though they prefer loose confederations of communities rather than states.

Environmental justice

The environmental justice movement in the United States has roots in 1978, when the Love Canal Homeowners' Association was organized by residents whose houses turned out to be on top of an abandoned toxic waste dump once operated by the Hooker Chemical Corporation in Buffalo, New York. After dragging its feet, the federal Environmental Protection Agency eventually agreed that this situation was unhealthy and dangerous, and the federal government bought out the residents' homes. Love Canal catapulted Lois Gibbs to stardom; previously an unpolitical working-class housewife, she went on to organize and head the national Citizens' Clearinghouse on Hazardous Waste (later renamed the Center for Health, Environment, and Justice). The other foundational action was a 1982 struggle against plans for a dump for toxic PCBs in Warren County, North Carolina, a predominantly African-American community. The dump was built, but the idea of 'environmental racism' began to achieve prominence.

The environmental justice movement is concerned with the degree to which the environmental risks generated by industrial society fall most heavily on the poor and ethnic minorities (Szasz, 1994). Issues of class and race, traditionally ignored by a mainstream US environmental movement composed mostly of middle-class whites, are highlighted, especially in the idea of 'environmental racism.' The risks in question related initially to toxic waste dumps, but concern soon broadened to encompass nuclear facilities, waste incinerators, air and water pollution, mining operations as they threatened the health of rural people (especially Native Americans), and pesticide use as it threatened the health of migrant farm workers. The movement grew out of thousands of groups organized locally to fight particular environmental threats. Derided by their opponents as having only a NIMBY (Not In My Back Yard) orientation, local groups soon got in touch with each other in conscious pursuit of a goal of NIABY (Not In Anybody's Back Yard). As Lois Gibbs puts it, the idea is to 'plug the toilet' on toxic wastes, and force industry to stop producing them in the first place (Dowie, 1995: 126). The movement therefore opposes the risk management paradigm discussed in Chapter 4, seeking instead to prevent the generation of risks (Bullard, 1999: 3).

The distinctive organizational form of the movement is the network

(Schlosberg, 1999). Local groups relate to each other without any national leadership or bureaucratic structure. This form is very different from that of the mainstream environmental groups, such as the National Wildlife Federation, Sierra Club, and Environmental Defense Fund, with their plush offices in Washington DC, highly paid chief executives, and easy access to the corridors of power.

The contrast with the mainstream groups is dramatic enough for environmental justice to be styled an alternative environmental movement. The networks of this second movement can bring together otherwise very different kinds of people: for example, white suburban housewives, inner city blacks, and Native Americans on reservations, united in opposition to a particular polluter or an interconnected set of environmental threats. Schlosberg (1999) stresses the 'critical pluralism' embodied in the movement, as groups with very different characteristics reach out across their differences in respectful fashion as they engage common adversaries. When it comes to tactics, the movement is eclectic. Like the mainstream, it engages in litigation and lobbying, but it is also more comfortable with confrontational tactics involving demonstrations, blockades, sit-ins, and boycotts.

Environmental justice activists can resent the fact that mainstream groups came late to toxics issues, especially as they relate to race and class, but then sought foundation funding for efforts in these areas. In some cases mainstream groups have combined late entry with negotiated deals with polluters that fall far short of ending the generation of toxic wastes. Lois Gibbs, for one, does not even like being called an environmentalist. As she puts it, environmentalists are people who eat yoghurt, while her people drink Budweiser and smoke (Dowie, 1995: 171).

With its conversion from NIMBY to NIABY strategies, environmental justice eventually raises the structural issue. The implication of 'plugging the toilet' is a transformed political economy, one in which hazardous wastes are no longer conceptualized as byproducts to be dealt with as an afterthought. Rather, these wastes are evidence of fundamental irrationality in the system, demanding cure in the form of production planning to eliminate the generation of wastes. Some clear parallels with ecological modernization can be discerned here. The main difference is that ecological modernizers believe capitalist enterprises themselves can seek efficiency and profit through waste minimization, whereas the environmental justice movement believes that such changes can only be forced upon reluctant corporations through political action.

Environmental justice is perhaps only weakly ecological: that is, there is little appreciation of the role played by complex ecosystems in sustaining life on earth. Yet this lack of appreciation is perhaps a contingent feature stemming from the movement's origins in existing community groups faced with some immediate local health hazards. The movement could probably only benefit from an appreciation of the ecosystemic nature of things, as this would provide further justification for the network form of social and political organization, and further support for the need radically to overhaul the industrial political economy.

Environmentalism of the global poor

Eco-socialism and environmental justice both have a special interest in the most materially disadvantaged members of society. The global impact of both movements remains somewhat limited. Guha (1997) refers to 'the environmentalism of the poor,' who experience degraded environments very directly as a result of deforestation, falling water tables, soil erosion, physical displacement by large projects such as dams, industrialized fishing, and high-technology agriculture. One of the first protests in reaction to such assaults was the Chipko movement against logging in India in the 1970s. Other actions have opposed dams, resource reallocation to the wealthy (such as timber companies operating in tropical forests), biopiracy by corporations that steal local ecological knowledge about valuable plant products (and sometimes even try to sell the knowledge back), the privatization of common land, and the creation of debt. The repertoire of such movements now includes nonviolent civil disobedience, hunger strikes, occupations, demonstrations, media publicity, and seeking support from international NGOs. In 2004, Kenyan environmentalist Wangari Maathai won the Nobel Peace Prize. Her Green Belt movement was responsible for planting more than 30 million trees since 1977. The idea is that reforestation fights soil erosion and provides fuel sources; the movement links environmental protection and social justice.

These movements and actions are preoccupied with immediate material needs and so at first sight appear a bit distant from more ecocentric greens in the developed world. Many movements in the Third World are not solely environmental, being linked to independence struggles (for example, in French possessions in the South Pacific), anti-corruption, political reform, and democracy (Haynes, 1999). To the extent that environmental degradation and economic damage flow from expansion of the global political economy into their land and lives, activists can make common cause with anti-globalization protestors. Both combine economic and environmental concerns (Shiva, 2000). If a Third World movement can gain the attention of international NGOs, then its cause may prosper. The Ogoni people in Nigeria, fighting both the Nigerian government and oil companies despoiling their land, have been especially successful in this respect, and the Shell corporation tried to change its ways after being embarrassed by the resulting international publicity. However, most struggles fail to make the international limelight, and the attention of NGOs is itself a scarce resource. Sometimes the 'environmentalism of the poor' may conflict with environmentalists whose idea of wilderness has no people in it. Indigenous people who have lived on and with this land find their interests sidelined (Bayet, 1994). Still worse, they may find themselves expelled from land designated as wilderness preserve. Particularly tragic is the case of Kenya, where Masai, Turkana, and Ndorobo people were expelled in the interests of elephant conservation in the early 1990s (Haynes, 1999: 230–2).

Anti-globalization and global justice

The anti-globalization movement came to public attention following street battles outside the 1999 meeting of the World Trade Organization (WTO) in Seattle. Since then, meetings of the WTO, World Economic Forum, International Monetary Fund, and G8 group of wealthy economies are normally accompanied by protests, except when they are held in places like Qatar where protestors are not admitted. The protests are the visible manifestations of a larger network of discontent. Unlike some more established social movements, the protestors do not really have a coherent program, still less an alternative to global capitalism. Even the term 'anti-globalization' is really just journalistic shorthand, an over-simplification given that protest itself is globalized. The protests were initially ridiculed by the global economic establishment and its supportive media, but this establishment has been forced to notice and address a range of concerns it had previously ignored under its drive for free trade and capital mobility. The particular concerns of the protestors have included the political power of multinational corporations, the effect of marketization on the global environment, lax labor practices such as sweatshops and child labor, loss of jobs in high-wage societies, and unfair terms of trade between rich and poor countries. Only some of the concerns have an environmental aspect. With time, the various concerns begin to coalesce into a more coherent critique of global capitalism.

Given the fluidity and variety it embodies, the movement can be interpreted in several different ways. Organizationally, it resembles the kind of network pioneered earlier by the environmental justice movement, taking networks and associated respect for diverse identities and positions across national boundaries. The connection to environmental justice is accentuated by the fact that many of the protestors prefer the title 'global justice' for their movement. The carnival-like atmosphere of the street protests resonates with Torgerson's (1999) proposals for a green public sphere that downplays the goal-oriented logic of environmental politics and its social movements. Alternatively, the movement could be interpreted in more moderate terms as signaling the need to bring transnational capitalism and its institutions (such as the WTO) under the control of national governments—and indirectly all the social interests that government is supposed to serve.

Animal liberation

With time, we have seen a gradual expansion in the range of human beings deemed worthy of a full range of rights: the poor, women, non-whites, children, disabled people, gays, and lesbians. Animal liberationists ask: why stop there? Why not extend the same rights to animals as well (see, for example, Regan, 1983)? The rights in question would involve rights against being killed for pleasure, against being used as food, against being imprisoned, against being experimented upon—even when such activities yield clear benefits to humans (for example, the medical benefits of animal experimentation). The leading animal liberationist Peter Singer (1975) argues that the criterion of sentience should be our guide (and not just for the gastronomic reason that it allows him to eat oysters).

Animal liberation is a movement as well as a philosophy (see Garner,

1993). Some of its more radical actions have involved the freeing of animals from factory farms, the destruction of animal experimentation laboratories, and firebombing of shops selling animal products. Britain's Animal Liberation Front is especially active. But animal liberationists are also active in more conventional pressure group and party politics, in part by trying to radicalize long-established animal welfare organizations such as Britain's Royal Society for the Prevention of Cruelty to Animals. Peter Singer himself was a Green Party candidate for the Australian Senate in 1996.

Singer's candidacy notwithstanding, animal liberation sometimes fits uneasily in green discourse because it is weakly ecological, some would say anti-ecological. For in its concern with individual creatures, it can lose sight of larger ecological connections. Does not the wellbeing of the ecosystem sometimes require the deaths of individual creatures (for example, the elimination of exotic species such as cats and foxes, which in Australia are wiping out native species)? What are we to say about predation in the natural world? Should predators be forced to become vegetarian, on the grounds of their violation of the right to life of their prey? Are not humans, as many deep ecologists insist, 'naturally' hunters? Animal liberationists might connect to green concerns because a commitment to the rights of wild animals means a commitment to preservation of their habitat. However, in England, countryside has in places been preserved against industrial agriculture for the sake of providing habitat for foxes—and fox hunting. Similarly, in Spain, bullfighting requires preservation of wild areas in which bulls are raised and roam.

Perhaps a more secure connection to ecological concerns would rest on the fact that an end to the exploitation of animals for human purposes raises all kinds of structural political-economic questions. It would mean a thorough reorientation of agriculture and agribusiness, and in power relationships more generally—the power of humans to use land and other species for their own benefit (Benton, 1993).

Discourse analysis of green politics

The storyline of radical green politics points to multifaceted social and ecological crises that can only be resolved through political action and structural change. Alternative forms of consciousness may be welcome as part of this project, but both the causes of crisis and the required approaches to solution involve a lot more. Complex social relations are at issue too, and action needs to take place both within and upon these relationships.

Basic entities whose existence is recognized or constructed

Green radicalism's urgency in the face of crisis is backed by recognition of ecological limits (though in eco-Marxism such limits are treated as applying only to capitalism, and in environmental justice and the environmentalism of the global poor limits receive little explicit attention). Nature is recognized in the form of complex ecosystems whose wellbeing requires that humans change their ways. But the necessary change is not simply cultural. Green politics emphasizes reflection and reasoning, though this does not mean that humans have to be *homo economicus* individuals, concerned only with calculation of what is in their own immediate material interest. Human horizons can and should be much wider. In contrast to a more cultural approach, social, political, and economic structures are recognized as having an important influence that cannot be reduced to the sensibilities of the individuals inhabiting them.

Assumptions about natural relationships

Green politics assumes a natural relationship of equality across individuals, at least in terms of the capacity to engage in reasoned communication about collective ends. Hierarchy that both pre-dates and is reinforced by modernity is recognized and condemned. The most elaborate analysis of hierarchy comes in social ecology, whose political philosophy is rooted in a demonstration of the unnaturalness of hierarchy of all kinds.

Despite its core egalitarianism, green politics allows compromise with other kinds of relationships, such as competitive ones, especially in its contemplation of economic systems. Such competitive relationships should, however, be kept in check by more egalitarian political structures. The precise character of desirable political structures is disputed. There is a substantial gap between the quasi-anarchism of social ecology and the statism of *Realo* greens and eco-socialists.

When it comes to specifying appropriate relationships linking human systems and natural systems, there is (or can be) a strong conception of complex ecological connections. Unlike deep ecology this conception does not have to reduce to any simple biocentric egalitarianism. Humans can be set apart from nature by virtue of their reasoning capacities, but this does not warrant hierarchy and domination of nature. A stewardship relationship is more likely to be posited.

Agents and their motives

Political agency is granted to a variety of actors, both individual and collective, and so encompasses movements, parties, and states as well as persons. Collective actors are central to this side of green politics, in contrast to the more individualistic emphasis in green consciousness change. The possibility that there might be agency in nature is generally downplayed, except in social ecology (see also Dryzek, 1990*a*).

Green radical politics is likely to treat the essence of human motivation as multidimensional, at once competitive and cooperative, violent and peaceful, instrumental and communicative, selfish and public-spirited. Political life is mainly about promoting institutional structure and political action to evoke the more benign motivation in each of these pairs, and control the more nefarious one.

Key metaphors and other rhetorical devices

Over the last few hundred years, the modern world has been constructed in terms of mechanistic images of both human social systems and natural systems (as we have seen in earlier discussions of Promethean discourse and economic rationalism). Greens reject this imagery. Metaphors present in green radical politics are perhaps less vivid and colorful than those featured in green cultural change, though there is a shared emphasis on organic metaphors. The world is approached in terms of organic balance, where wholes cannot be understood by reduction to their component parts, and living things interact in ways that can never be understood fully. Human systems may be irrational at present (committed, for example, to blind pursuit of material riches, or the generation of toxic wastes with nowhere to put them), but they are capable of a greater rationality in their interactions with natural systems. We can apply standards of ecological rationality to the analysis and redesign of these systems. Social systems, like individuals, must be treated as capable of learning.

Green politics involves argument, not just appeal to the emotions. The accompanying rhetoric is likely to appeal to ideals of progress beyond an irrational industrial order, rather than promise return to some primal Eden. Like sustainable development and ecological modernization, a belief in progress is grounded in a model of individual human development.

Green politics in practice

Green political action has sought to change institutions, practices, and policies. Its impact should be sought not just in the tangible achievements of particular parties, networks, or other green organizations, but also in the degree to which green discourse has permeated political-economic life more generally.

Green parties have been represented in the parliaments of an increasing number of countries since 1981, when the Francophone Ecolo and Flemish Agalev parties won seats in the Belgian parliament. The highest vote achieved by greens in any parliamentary election is 18.2 per cent in the 2002 state election in Tasmania (Australia). The highest vote of any green

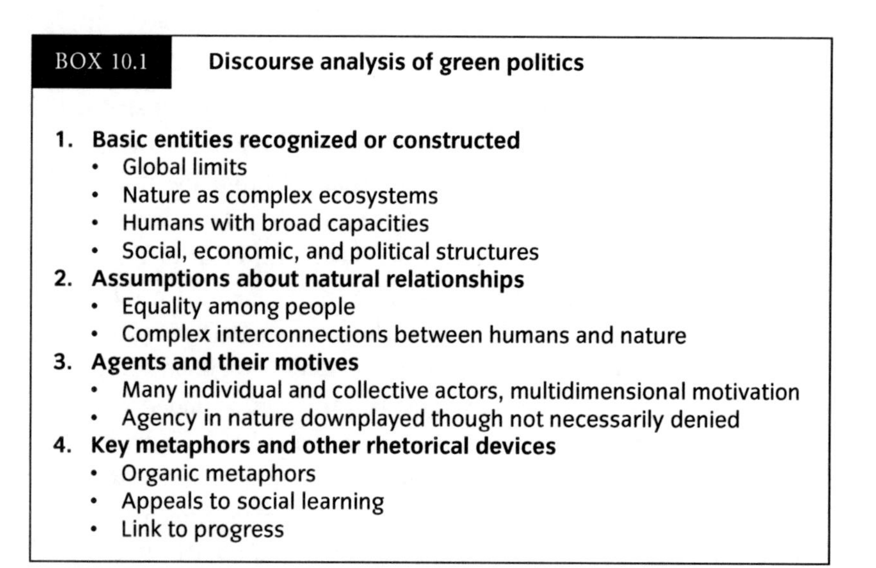

BOX 10.1 **Discourse analysis of green politics**

1. **Basic entities recognized or constructed**
   - Global limits
   - Nature as complex ecosystems
   - Humans with broad capacities
   - Social, economic, and political structures
2. **Assumptions about natural relationships**
   - Equality among people
   - Complex interconnections between humans and nature
3. **Agents and their motives**
   - Many individual and collective actors, multidimensional motivation
   - Agency in nature downplayed though not necessarily denied
4. **Key metaphors and other rhetorical devices**
   - Organic metaphors
   - Appeals to social learning
   - Link to progress

party in a nationwide election was achieved in 1989 by the British Greens in elections to the European Parliament. But their 15 per cent of the vote won them a grand total of zero seats, and they have yet to come anywhere close to winning a seat in the Westminster parliament, though in 1992 Cynog Dafis, of the Welsh nationalist party Plaid Cymru, was elected with green backing. By 1996 the British Greens were debating at their annual conference whether it was even worthwhile contesting the next general election. Their electoral difficulties illustrate the degree to which the success of green parties in winning seats depends crucially on the kind of electoral system in operation. The first-past-the-post or simple plurality system disadvantages small parties and hinders their growth, explaining the minor electoral impact of green parties in the United States and United Kingdom. This system was the norm in the English-speaking world, though even here proportional representation can be found in the Irish Republic, the Welsh and Scottish assemblies, upper houses in Australian parliaments, and, beginning in 1996, the parliament of New Zealand, in all of which green parties have secured seats. Proportional representation advantages all small and emerging parties, not just green ones. Green parliamentary representation has long been important in the German *Bundestag*, and in 1998 *Die Grünen* joined a federal governing coalition for the first time. Their leader, Joschka Fischer, became Foreign Minister, indicative of the degree to which the party's leadership now downplayed the radical transformation of society.

The trajectory of green votes and green seats reveals a spread in the number of countries with electorally significant green parties, but no takeoff in the strength of any one party beyond single percentage points in terms of votes or seats. This is surprising on the face of it, given the degree to which voters in post-industrial societies should in principle be sympathetic to many of the values that green parties stand for (see Inglehart, 1990).

But the real impact of green parties may be in the degree to which they have forced more established 'grey' parties, and the political system as a whole, to craft responses to the green electoral threat. The development of the discourse of ecological modernization detailed in Chapter 8 can be interpreted as an attempt by the prevailing political order to head off the green challenge. Ecological modernization has developed in the European heartland of green party politics, and has appropriated more than a few ideas originally developed by the greens. Certainly, ecological modernization lacks the radical edge of green politics. Yet it still posits a structural transformation of capitalism. The irony is that if this transformation succeeds it will deprive green radicalism of its bite by showing that transition to a totally different political economy is unnecessary. The historical analogy here is with the rise of socialism, which in the early to mid-twentieth century forced the capitalist political economy to develop welfare states and full employment practices, thus blunting the radical edge of the socialist critique of capitalism. The difference is that socialist parties could often dominate parliamentary majorities, and sometimes formed governments by themselves.

Green parties have, however, participated in governing coalitions in city, state, and national governments in several European countries, normally alongside parties of the social democratic left. In the Italian national election of 1996 the greens formed part of the victorious Olive Tree alliance, and green parties have joined national governments in Belgium, Finland, and France, as well as Germany. Governments containing greens have not always enacted policies radically different from social democratic governments that do not contain greens, and no program of wholesale political-economic transformation has yet come from any such coalition. The price of participation in government has often been a heavy one in terms of moderation of green demands. Green parties can rarely make a credible threat of defection to a coalition with larger right-wing parties, and so their social-democratic coalition partners can sometimes take their support for granted (Poguntke, 2002: 138).

The most significant achievement of a green party in power may be the planned phase-out of nuclear power secured by the German Greens after 1998. However, this plan was to take effect over several decades, so it could be reversed by a subsequent government. Its terms were negotiated between economics ministry officials and nuclear industry executives (Dryzek et al., 2003: 189). The German Greens were actually quite successful in preventing the construction of nuclear installations in their prior days as an oppositional social movement (for example, in forcing the cancellation of a reprocessing plant at Wackensdorf in 1989).

Yet to focus on any seeming lack of policy impact by green parties in government would be to miss the crucial role of green discourse in transforming the terms of political debate, and requiring other parties to adjust their positions on environmental issues and other green concerns. The German Greens have a name for this: *themenklau*, the stealing of green ideas by grey parties.

Not everyone within the green movement believes that electoral politics is the proper focus of green energies. Political life is not just party politics. It can also cover discussions in bars and coffee shops, community organizing, educational efforts, self-help groups, boycotts, demonstrations, strikes, blockades, ecotage, sit-ins, and various kinds of media events. Greenpeace specializes in media events—for example, inserting a plug into the end of a pipeline discharging radioactive waste into the Irish Sea from the British Nuclear Fuels Ltd installation at Sellafield, pulling up genetically modified crops, and obstructing whaling vessels. With time, the green direct action repertoire expands. In Britain, Reclaim the Streets developed the tactic of shutting down traffic on urban streets, organizing by word of mouth and the internet in order to beat the police to a site. In the British countryside, tree-sitters and tunnelers put themselves in the way of motorway construction—winning the public relations war even as each road was eventually constructed (Doherty, 1999). To avoid the grasp of government and its compliant legal system, these British activists did not organize formally. These unorganized groups were unfettered in their activism, and did not have to worry about their assets being seized or their access to government threatened. Such worries frightened off more established groups such as Friends of the Earth, which looked uncomfortable: their heart was with the protestors, but expediency meant they had to keep their distance (Rootes, 2003: 5–6).

In Germany, anti-nuclear protests on a large scale continued after the Greens joined the federal governing coalition. In 2001 these protests targeted shipments of nuclear wastes for reprocessing—after Green Party leadership had acquiesced in the shipments as part of the price for securing the planned nuclear phase-out. Confrontations occurred between Green politicians and green protestors.

Shunning the state in favor of movement politics might seem to some an abdication of ambition, even of responsibility, leading to voluntary exile in a political wasteland. But such a perception is in error. Political pressure can be exerted at a distance upon the state. Here, social movements have at their disposal a number of instruments. They include the rhetorical ability to change the terms of policy debate, creation of fear of political instability, the production of ideas, and the embarrassment of governments. Much of the success of the anti-globalization/global justice movement since 1999 can be interpreted in these terms. The movement has forced social justice and environmental issues onto the agenda of economic organizations such as the WTO and IMF.

When it comes to changing public ideas and attitudes, it is hard to disentangle the relative influence of more mainstream environmentalism and green radicalism, but at least in Europe the established mainstream environmental groups (such as the Council for the Protection of Rural England, the World Wildlife Fund, and *Deutsch Naturschutzring*) have been rather staid and unimaginative. They were around long before the 1970s upsurge in green politics, with very little to show in terms of value change in society at large. It is the green radicals who have made the running in instigating change in ideas and attitudes, which extends to those who do not vote for green parties, still less join more radical green actions. Aspects of this change include (dim) awareness of ecological limits, sensitivity to the risks generated by industrial society in terms of chemical, nuclear, and biotechnological hazards, and of the possibilities for a more convivial way of life than the aggressively individualistic materialism of contemporary market society. Green radicalism has had less success in achieving broad acceptance of its core values relating to grassroots democracy and structural change in the political economy.

Green politics itself helps to constitute a parallel political society where at least some individuals can lead their social and political lives, an alternative to the grey mainstream of party politics (see Dryzek, 1996*a*: 46–53). This 'green public sphere' acts as a standing reminder to industrial society of the error of its ways, a place where critiques can be generated and alternatives explored (Torgerson, 1999). This parallel polity can be oriented to the public policy debates of the day, even as a critical distance from mainstream politics is maintained. This kind of oppositional sphere was long the hallmark of green politics in Germany, especially when environmentalists lacked any points of access to government. Lack of access did not necessarily mean lack of influence, especially when it came to blocking environmentally destructive projects such as nuclear plants. And when the German government did open up in the 1990s, activists who had honed their skills and critiques in a radical public sphere sometimes succeeded in getting critical concerns onto the policy agenda, if not necessarily into policy practice (Dryzek et al., 2003: 190–1).

Green politics can also involve action oriented to the solution of particular well-defined problems in a fashion that attempts to reclaim political authority from the state. For example, in 1995 Greenpeace activists occupied the Brent Spar, an oil storage platform whose working life in the North Sea had come to an end. The Shell corporation intended to dispose of the platform by towing it into deeper water in the North Atlantic and sinking it. The publicity generated by Greenpeace, which also organized a consumer boycott of Shell throughout Europe, forced the company to change its plans and dispose of the platform on land. Shell's decision angered the British government, which was prepared to use force to evict the Greenpeace protestors. In this case at least, green activists possessed more political authority in relation to Shell than did the British government. In another case from the early 1990s, Greenpeace succeeded in persuading German paper companies to stop using chlorine to bleach paper. The key point in the campaign came when Greenpeace printed a plagiarized edition of *Der Spiegel*, the leading news weekly, on paper bleached without chlorine, with a hardly noticeable difference in quality.

Another case of such reclamation of political authority arises with the practice of 'popular epidemiology,' or community-based research on risk assessment. The paradigm case occurred in the community of Woburn in Massachusetts, where citizens angered by government denial that any problem existed organized a group called FACE (For a Clean Environment) which then conducted its own survey of the incidence of leukemia and birth defects which its members believed were linked to toxic waste sites. State and federal government agencies rejected this effort, arguing that the citizen risk investigators had no proper training in risk assessment, such that their findings were unreliable. Yet the results assembled by FACE were used as evidence in a lawsuit which was eventually settled out of court by one of the companies that had dumped toxics (see Brown and Mikkelsen, 1990). This case formed the basis for the 1998 film *A Civil Action*, starring John Travolta.

In the United States, tangible impacts are associated with the environmental justice movement (under which the Woburn action can also be classified). Unlike green parties, this movement has developed without contemplation of any kind of political or social theory, and without any debate over the appropriate degree of engagement with conventional political action and the state (though debate has occurred over how to relate to established mainstream environmental groups). Instead, the movement has moved on a variety of fronts, involving conventional litigation and lobbying as well as demonstrations, blockades, and boycotts. In the negative, the movement has achieved many victories, blocking plans for noxious facilities and forcing corporations and governments to compensate victims. No significant toxic waste disposal facility has been constructed in the United States since 1980. The toilet is well and truly plugged. As yet, this plugging has not led to any real movement toward a greener economy. However, the simple presence of environmental justice and its network form of organization is itself a significant political development, pointing to a kind of politics that is more authentically democratic and more green (in terms of green rationalism) than its mainstream alternative.

Environmental justice has permeated the highest levels of policy making in the United States, forcing at least a symbolic response from President Clinton in signing an executive order in 1993, declaring that federal environmental agencies must henceforth take principles of environmental justice into account in their decision making. Institutionalization came in the form of the Office of Environmental Justice and the National Environmental Justice Advisory Committee, both associated with the Federal EPA, though question marks remain about the effectiveness of both.

The environmentalism of the global poor is highly varied in its content and strategies, and so its impact defies easy summary. There have been local successes in (for example) forcing oil companies to clean up their operations, preventing the construction of large dams, and slowing deforestation. More positive projects include experiments in agro-ecology and community control of resources. But structural change in the political economy generally remains a remote prospect, and the reach of the globalized political economy continues to grow.

If environmental justice and anti-globalization represent achievement in practice without much in the way of theoretical reflection, social ecology and eco-socialism represent intellectual achievements without much obvious accompanying political or economic practice.

Being green in global capitalist times

In the last three decades or so green radicalism has come from nowhere to develop a comprehensive critique of the environmental, social, political, and economic shortcomings of industrial society. As such, it represents perhaps the most significant ideological development of the late twentieth century. Yet there remains a great deal of uncertainty about the best way to practice green politics in the face of a seemingly recalcitrant and secure liberal capitalist political economy, increasingly entrenched at the global level beyond the control of most national governments. This political economy is reinforced by several of the discourses analyzed in earlier chapters. Prometheans of course have nothing but scorn for green critiques, and when they do address green thinking completely trivialize it. The three problem-solving discourses see no need for the kind of wholesale structural change green rationalists seek. Sustainable development and ecological modernization are more likely to take these concerns seriously, but believe a structural response can be crafted that does not involve abandoning the basic parameters of liberal capitalism, even as they are uneasy with the global spread of market liberalism.

So just what kind of alternative political economy do greens want? Is capitalism to be overthrown, transcended, or transformed? Certainly, capitalism as it currently stands is regarded as unacceptable, but most greens remain uncertain about exactly what to do with it. Even eco-Marxists and eco-socialists are a bit coy about whether they envisage capitalism being overthrown in favor of some socialist alternative, or even gradually transformed into such an alternative. These viewpoints perhaps make more sense as critical devices, in which socialism is raised mainly as a way of highlighting the flaws of capitalism, the real-world prospects for any kind of socialism (be it reformist or revolutionary) having receded rather dramatically.

But to demand a blueprint for an alternative society may be asking too much. If the twentieth century holds one political lesson, it is that we should beware of anyone peddling such blueprints, be they socialist paradises, fascist Reichs to last a thousand years, or free-market utopias popularized in the Anglo-American world in the 1980s and exported in the form of 'shock therapy' to several East European countries after 1989.

Whatever the leanings of their advocates and supporters, such blueprints inevitably go wrong when confronted with the complexities of the real world, and bring at best only the kind of state centralization and authoritarianism introduced in Britain by Margaret Thatcher (ironically alongside a rhetoric of freedom and choice). At worst, they bring totalitarianism and a police state. The explanation is simple: as soon as real world surprises come along, proponents of the blueprint feel they have to save it via increasingly coercive measures. The notion that the blueprint itself may be flawed never crosses their minds (for more detail on this general argument, see Popper, 1966, 1972; for an application to free-market utopias in Eastern Europe, see Pickel, 1993).

In this light, the fact that greens do not have any well-defined blueprint for a new society twinned with a coordinated strategy for achieving it is actually a point in their favor. What green rationalists do have in abundance are ideas that can be pressed into a decentered approach to the achievement of a greener society, where there is room for a variety of experiments whose general orientation is given by green discourse, but whose specifics can vary quite substantially. Such variety is the essence of the green public sphere (Torgerson, 1999). Bioregional projects, networks of community activists, oppositional political fora, experiments in local grassroots democracy, social ecology's radical municipalism, and attempts to radicalize democratic pragmatist initiatives of the sort discussed in Chapter 5 can all fit in here.

Such a decentralized approach fits quite nicely with green ideas about local initiative and community self-control. However, can such a loosely coordinated set of responses ever be adequate in the face of a liberal capitalist political economy which is more secure and more entrenched than ever before in history? This system is increasingly geared to free trade, economic growth, and the mobility of investment capital across national boundaries. It is this system which is the dominant political reality of our times. All national, regional and local governments now see it as their first task to accommodate themselves to the imperatives of this system, to keep investors happy by promoting a positive climate for business. A decentered program of green initiatives might appear just a minor set of irritants to this monolithic, global, transnational capitalist political economy. If in the face of this behemoth green politics is to be more than theatre, then perhaps it has something to learn from other discourses of environmental concern. Some possibilities along these lines are addressed in the concluding chapter.

NOTE

1 Whose pages I have also graced (Dryzek, 1992*c*).

PART VI
CONCLUSION

11

Ecological Democracy

What can be said by way of conclusion about how the various discourses have survived the questions asked of them, and their comparison with other discourses? First, there are some complementarities across discourses. For example, a weak form of ecological modernization is quite compatible with administrative rationalism's strong state and some of the instruments advocated in economic rationalism (such as green taxes). And green radicalism is happy to accept the basic idea of global limits developed by survivalists—though not survivalists' political analysis and prescription. Equally clearly, there can be plenty of tension between discourses. Survivalists have core disagreements with Prometheans, sustainable development, and ecological modernization. Economic rationalists are never going to agree with administrative rationalists, democratic pragmatists, or green radicals about the best way of ordering environmental affairs.

One way of easing the tensions somewhat is to note that different discourses may be applicable to different kinds of problems. In essence, survivalism and Promethean discourse are about global issues. Whatever position one reaches in the dispute between them, it would be possible to follow any one of the three problem-solving discourses at the local level (though Prometheans might say that even such local efforts are often unnecessary). Other compartmentalizations might be found: for example, one might be green when it comes to lifestyle, but a democratic pragmatist when it comes to policy.

Such reconciliation possibilities notwithstanding, it remains the case that each of the discourses analyzed offers a reasonably comprehensive account of and orientation to environmental affairs at all levels, from the global to the local, and across different issue areas (pollution, resource depletion, wilderness protection, and so forth). This comprehensiveness certainly applies to Promethean discourse, administrative rationalism, democratic pragmatism, economic rationalism, sustainable development, and green radicalism. It is less applicable to survivalism, which concerns itself only with global affairs, and ecological modernization, which has so far addressed only how industrial economies might be restructured, with little application to non-industrial societies or global analysis.

With these competing comprehensive visions in mind, I would argue by way of approaching a conclusion that any intelligent approach to environmental issues demands two things. The first is a dynamic, structural-level analysis of the liberal capitalist political economy, where it might be headed, and what realistically can be done to alter this trajectory to more ecologically benign ends. For a confident and globally organized liberal capitalism mostly insensitive to environmental concerns is the dominant political fact of our times. So without such an analysis, we are reduced to wishful thinking about how things might be different. Of the discourses surveyed, ungrounded wishful thinking about a different world characterizes survivalism, economic rationalism, and aspects of gre-n radicalism—though of course they wish for very different things! Only two of the discourses provide a coherent analysis of the kind needed: Promethean discourse and ecological modernization.

Prometheans believe that the current trajectory of liberal capitalism is unproblematical, and that all we need do is leave it alone to provide abundance for humanity, in the future as in the past. Ecological modernizers, in contrast, recognize that laissez-faire liberal capitalism is environmentally destructive. Thus they seek an ecological restructuring of capitalism that respects the constraints imposed by this economic system on political action, and which is consistent with the basic imperatives of the system. If one accepts the Promethean viewpoint, then the matter ends. On the other hand, if one rejects that viewpoint—and I argued in Chapter 3 that there are good reasons to do so—then the second quality demanded by an intelligent approach to environmental affairs comes into play.

This second quality is the capacity to facilitate and engage in social learning in an ecological context. Environmental issues feature high degrees of uncertainty and complexity, which are magnified as ecological systems interact with social, economic, and political systems. Thus we need institutions and discourses which are capable of learning—not least about their own shortcomings. Survivalism, Promethean discourse, administrative rationalism, and economic rationalism provide few such resources, and exhibit little or no awareness of their own limits. In contrast, resources for this learning project are provided by democratic pragmatism, sustainable development, ecological modernization, and green radicalism. In each case, though, our appropriation from the discourse must be selective.

From democratic pragmatism come discursive procedures for the resolution of disputes through cooperative problem solving. Such procedures, including policy dialogue, environmental mediation, lay citizen deliberation, governance networks, and societal dialogues, are often limited in their scope and constrained by the structural context in which they operate. Critics of them rightly note that they can involve co-optation and neutralization of troublemakers by powerful government and corporate officials. The key, then, is try to break these shackles, moving such procedures in the direction of what I have described elsewhere as discursive designs, which transgress the boundaries of democratic pragmatism by pointing to a more radical democracy. Discursive designs involve collective decision making through authentic democratic discussion, open to all interests, under which political power, money, and strategizing do not determine outcomes (see Dryzek, 1990*b*: 29–56). A careful search reveals an ever-growing number of cases, some of which I have mentioned under democratic pragmatism and green radicalism (see also Smith, 2003).

From sustainable development comes the possibility of a decentered approach to the pursuit of sustainability. While at first glance the sheer variety of available definitions of sustainable development seems like a defect of this discourse, from the perspective of social learning it is a distinct advantage, for it does not rule out a variety of experiments in what sustainability can mean in different contexts, including the global context. A decentered approach to sustainability meshes quite nicely with discursive designs, which could find roles as the steering institutions and reflective components of experiments in sustainability—as Torgerson (1994, 1995) also recognizes.

From ecological modernization comes the possibility of a 'strong' or 'reflexive' version of the discourse, the essence of which goes beyond the retooling of the economy with waste reduction and profitability in mind. Ecological modernization so radicalized can involve institutional change in the direction of democratic experimentation, and open-ended exploration of what ecological modernization itself might mean. The very idea of reflexive development is that it is self-monitoring and critically aware of itself, thus conducive to social learning. Again, this fits quite nicely with a decentered approach to sustainability and discursive designs.

From green radicalism come the reasons why democratic pragmatism, sustainable development, and ecological modernization need to be radicalized to begin with. Green (radicalism) can bring to them a sense of urgency that survivalism shares but finds more difficult to disseminate, given that survivalism's imagery of certainty leaves little space for search and experimentation. Moreover, survivalism's flirtation with authoritarianism alienates democratic pragmatists, sustainable developers, and ecological modernizers alike. Green cultural change can also contribute here to the extent its adherents can become interested and involved in social institutions. Green radicalism can also remind us that oppositional politics in the public sphere and social movements can play a key role in social learning, which does not have to be tied to conventional politics (and may indeed proceed more readily outside the realm of conventional politics). Green politics can further bring to bear plenty of ideas about how political and economic institutions might look in an ecological future beyond industrial society. Linking these ideas to the other three discourses is a way of grounding such ideas in a more realistic analysis of how the future can actually unfold, as opposed to wishful thinking about how it should unfold.

The common thread that can be developed here is a renewed democratic politics, an ecological democracy.[1] But would such a politics indeed promote ecological values? One affirmative answer comes from democratic pragmatism: the kinds of values that can survive authentic democratic debate are those oriented to the interests of the community as a whole, rather than selfish interests within the community (or outside it). Foremost among such community interests is the integrity of the ecological base upon which the community depends. From green radicalism comes a reminder to democratic pragmatism that existing liberal democracies typically frustrate such processes: the influence of power, money, and strategy need to be unmasked and countered, as does the degree to which human communities have lost any sense of their ecological foundations.

For democracy, if it is about anything, is about authentic communication. Overcoming the impediments that distort such communication is crucial. One such impediment, ignored in the history of democratic theory but now exposed by the rise of green thinking, concerns communication with the nonhuman world. It would be absurd to think of that world as having preferences, or able to 'vote,' which is why most models of democracy are of limited applicability in a green context. But the nonhuman world can communicate, and human decision processes can be structured so as to listen to its communications more or less well. Large bureaucracies operating according to standard procedures insensitive to local ecological contexts fail this test; bioregional authorities governed by citizens with a thorough knowledge of local circumstances are likely to do much better.

Ecological democracy blurs the boundary between human social systems and natural systems. There is an additional sense in which ecological democracy is democracy without boundaries. Ecological problems and issues transcend established governmental jurisdictions, such that democratic exercises may need to be constituted in order to fit the size and scope of particular issues. When established authority in governmental jurisdictions is recalcitrant, then such forums may need to be constituted as oppositional democratic spheres. The impact of nongovernmental organizations in international politics (highlighted in the discussion of sustainable development) can be understood in these terms. When it comes to politics above local action, the appropriate organizational form may often be the network, as developed by the environmental justice and anti-globlization movements discussed in Chapter 10.

This sort of democracy without boundaries is clearly very different from the institutions established by and in industrial society whose priorities still dominate today's world. However, occasionally crises can enable environmental concerns to come into alignment with those concerns, and so advances can be made in terms of both the democratization and greening of dominant institutions. Examples here would include the legitimation crisis faced by the Nixon administration around 1970, which was defused by reaching out to environmentalists as the least radical element of the counter-culture, and led to a burst of environmental legislation as well as inclusion of (moderate) environmentalists in government. More recent risk-related crises in Europe related to food safety (mad cow disease and genetically modified organisms) have had less thoroughgoing but still significant impact.

Ecological modernization allows a more persistent link with the key economic priority of governments. At the national level, the most prominent political configuration that currently looks at all feasible would feature a consensual corporatist state pursuing ecological modernization confronted by a lively green public sphere. Germany constitutes the best example of such a combination—though of course Germany is far from being an ecolgical democracy. One hazard accompanying the inclusion of greens and environmentalists in corporatist government is depletion of the public sphere, as former activists are attracted into government, and accept moderation and moderation as the price to be paid. This situation characterizes Norway, a top performer in cross-national environmental comparisons, but with no social movements or oppositional public sphere to push the country any further (Dryzek et al., 2003).

In less structurally hospitable contexts, such as the international system, Third World countries, or states under sway of market liberalism, the pursuit of ecological democracy is more difficult, but not futile. For discourses, including environmental ones, help to constitute and reconstitute the world just as surely as do formal institutions or material economic forces. And in this discursive realm, as we have seen, the beginnings of ecological democracy are already present. Environmentalism already flourishes in opposition to industrialism; but much remains to be done if industrial society is ever to give way to ecological society.

NOTE

1 More extensive discussion of ecological democracy may be found in some of my other writings (Dryzek, 1987, 1990*a*, 1992*c*, 1996*c*, 1996*d*).

REFERENCES

ABBEY, EDWARD (1975), *The Monkey Wrench Gang*. Philadelphia, Pa.: J. B. Lippincott.

—— (1990), *Hayduke Lives!* Boston, Mass.: Little, Brown.

ACKERMAN, BRUCE A. and HASSLER, WILLIAM T. (1981), *Clean Coal, Dirty Air: or How the Clean Air Act became a Multibillion-Dollar Bail-Out for High-Sulfur Coal Producers and What Should Be Done About It*. New Haven, Conn.: Yale University Press.

ALEXANDER, CHRISTOPHER (1964), *Notes on the Synthesis of Form*. Cambridge, Mass.: Harvard University Press.

AMY, DOUGLAS J. (1987), *The Politics of Environmental Mediation*. New York: Columbia University Press.

—— (1990), 'Decision Techniques for Environmental Policy: A Critique,' pp. 59–79 in Robert Paehlke and Douglas Torgerson (eds.), *Managing Leviathan: Environmental Politics and the Administrative State*. Peterborough, Ontario: Broadview.

ANDERSEN, MIKAEL SKOU (1994), *Governance by Green Taxes: Making Pollution Prevention Pay*. Manchester: Manchester University Press.

ANDERSON, FREDERICK L., KNEESE, ALLEN V., REED, P. D., STEVENSON, R. B., and TAYLOR, S. (1977), *Environmental Improvement Through Economic Incentives*. Baltimore, Md.: Johns Hopkins University Press for Resources for the Future.

ANDERSON, TERRY L., and LEAL, DONALD R. (1991), *Free Market Environmentalism*. Boulder, Colo.: Westview.

——, —— (2001), *Free Market Environmentalism*, revised edition. New York: Palgrave.

ARIAS-MALDONADO, MANUEL (2000), 'The Democratisation of Sustainability: The Search for a Green Democratic Model,' *Environmental Politics*, 9 (4): 43–58.

ARROW, K., COSTANZA, R., DASGUPTA, P., FOLKE, C., HOLLING, C. S., JANSSON, B. E., LEVIN, S., MALER, K.-G., PERRINGS, C., and PIMENTAL, D. (1995), 'Economic Growth, Carrying Capacity and the Environment,' *Science*, 268: 520–1.

Bailey, Ronald (ed.) (2000), *Earth Report 2000: Revisiting the True State of the Planet*. New York: McGraw Hill.

BARNET, RICHARD J. (1980), *The Lean Years: Politics in the Age of Scarcity*. New York: Simon and Schuster.

BARNETT, HAROLD J., and MORSE, CHANDLER (1963), *Scarcity and Growth: The Economics of Natural Resource Availability*. Baltimore, Md.: Johns Hopkins University Press for Resources for the Future.

BARRETT, BRENDAN, and FISHER, DANA (2005), *Ecological Modernization in Japan*. London: Routledge.

BARRY, JOHN (2003), 'Ecological Modernisation,' pp. 191–213 in Edward Page and John Proops (eds.), *Environmental Thought*. Cheltenham: Edward Elgar.

BARTLETT, ROBERT V. (1990), 'Ecological Reason in Administration: Environmental Impact Assessment and

Administrative Theory,' pp. 81–96 in Robert Paehlke and Douglas Torgerson (eds.), *Managing Leviathan: Environmental Politics and the Administrative State*. Peterborough, Ontario: Broadview.

BAYET, FABIENNE (1994), 'Overturning the Doctrine: Indigenous People and Wilderness—Being Aboriginal in the Environmental Movement,' *Social Alternatives*, 13 (2): 27–33.

BECK, HANNO, DUNKEN, BRIAN, and KRIPKE, GAWAIN (1998), *Citizens' Guide to Environmental Tax Shifting*. Washington, DC: Friends of the Earth.

BECK, ULRICH (1992), *Risk Society: Towards a New Modernity*. London: Sage.

—— (1999), *World Risk Society*. Cambridge: Polity.

——, GIDDENS, ANTHONY, and LASH, SCOTT (1994), *Reflexive Modernization: Politics, Tradition and Aesthetics in the Modern Social Order*. Cambridge: Polity.

BECKERMAN, WILFRED (1974), *In Defence of Economic Growth*. London: Cape.

—— (1995), *Small is Stupid: Blowing the Whistle on the Greens*. London: Duckworth.

—— (2002), *A Poverty of Reason: Sustainable Development and Economic Growth*. Oakland, CA: The Independent Institute.

BEDER, SHARON (2001), 'Neoliberal Think Tanks and Free Market Environmentalism,' *Environmental Politics*, 10 (2): 128–33.

BENNETT, JANE, and CHALOUPKA, WILLIAM (eds.) (1993), *In the Nature of Things: Language, Politics, and the Environment*. Minneapolis: University of Minnesota Press.

BENTON, TED (1993), *Natural Relations: Ecology, Animal Rights, and Social Justice*. London: Verso.

BEREJIKIAN, JEFFREY (1995), 'The Gains Debate: Framing State Choice.' Unpublished PhD dissertation, University of Oregon.

BERGER, THOMAS (1977), *Northern Frontier, Northern Homeland: Report of the MacKenzie Valley Pipeline Inquiry*. Toronto: James Lorimer.

BHASKAR, ROY (1975), *A Realist Theory of Science*. Brighton: Harvester

BIEHL, JANET (1991), *Rethinking Ecofeminist Politics*. Boston, Mass.: South End Press.

BLOWERS, ANDREW (1997), 'Environmental Policy: Ecological Modernisation or the Risk Society?' *Urban Studies*, 34: 845–71.

BOBROW, DAVIS B., and DRYZEK, JOHN S. (1987), *Policy Analysis by Design*. Pittsburgh, Pa.: University of Pittsburgh Press.

BOOKCHIN, MURRAY (1982), *The Ecology of Freedom: The Emergence and Dissolution of Hierarchy*. Palo Alto, Calif.: Cheshire.

—— (1986), *The Modern Crisis*. Philadelphia, Pa.: New Society.

—— (1990), *Remaking Society: Pathways to a Green Future*. Boston, Mass.: South End Press.

—— (1995), *Re-Enchanting Humanity: A Defense of the Human Spirit Against Antihumanism, Misanthropy, and Primitivism*. London: Cassell.

—— and FOREMAN, DAVE (1991), *Defending the Earth*. Boston, Mass.: South End Press.

BOULDING, KENNETH R. (1966), 'The Economics of the Coming Spaceship Earth,' in Henry Jarrett (ed.), *Environmental Quality in a Growing Economy*. Baltimore, Md.: Johns Hopkins University Press.

BOWERS, CHET (1999), 'The Role of Education and Ideology in the

Transition from a Modern to a More Bioregionally-Oriented Culture,' pp. 199–204 in Michael Vincent McGinnis (ed.), *Bioregionalism.* New York: Routledge.

Bradley, Robert L. (2003), *Climate Alarmism Reconsidered.* London: Institute of Economic Affairs.

Braithwaite, John, and Drahos, Peter (2000), *Global Business Regulation.* Cambridge: Cambridge University Press.

Bramwell, Anna (1989), *Ecology in the Twentieth Century: A History.* Cambridge: Cambridge University Press.

Brechin, Stephen R., Wilshusen, Peter R., Fortwangler, Crystal L., and West, Patrick C. (2003), *Contested Nature: Promoting International Biodiversity with Social Justice in the Twenty-first Century.* Albany: State University of New York Press.

Brooks, D. B. (1992), 'The Challenge of Sustainability: Is Integrating Environment and Economy Enough?,' *Policy Sciences,* 26: 401–8.

Brown, Lester R. (1978), *The Twenty-Ninth Day: Accommodating Human Needs and Numbers to the Earth's Resources.* New York: Norton.

—— (1981), *Building a Sustainable Society.* New York: Norton.

—— (2003), *Plan B: Rescuing a Planet Under Stress and a Civilization in Trouble.* Washington, DC: Earth Policy Institute.

——, Flavin, Christopher, and Postel, Sandra (1992), *Saving the Planet: How to Shape an Environmentally Sustainable Global Economy.* London: Earthscan.

Brown, Phil, and Mikkelsen, Edwin J. (1990), *No Safe Place: Toxic Waste, Leukemia, and Community Action.* Berkeley, Calif.: University of California Press.

Bryner, Gary C. (2000), 'The United States: "Sorry—Not Our Problem",' pp. 273–302 in William M. Lafferty and James Meadowcroft (eds.), *Implementing Sustainable Development: Strategies and Initiatives from High Consumption Societies.* Oxford: Oxford University Press.

Bullard, Robert (1999), *Environmental Justice in the 21st Century.* Online at **www.ejrc.cau.edu/ejinthe21century.htm**.

Burke, Tom (2001), *Ten Pinches of Salt: A Reply to Bjørn Lomborg.* London: Green Alliance.

Caldwell, Lynton K. (1978), 'The Environmental Impact Statement: A Misused Tool,' in Ravinder Jain and Bruce Hutchings (eds.), *Environmental Impact Analysis.* Urbana: University of Illinois Press.

—— (1982), *Science and the National Environmental Policy Act: Redirecting Policy Through Administrative Reform.* Tuscaloosa: University of Alabama Press.

—— (1984), 'The World Environment: Reversing US Policy Commitments,' pp. 319–38 in Norman J. Vig and Michael E. Kraft (eds.), *Environmental Policy in the 1980s: Reagan's New Agenda.* Washington, DC: Congressional Quarterly Press.

Capra, Fritjof, and Spretnack, Charlene (1984), *Green Politics: The Global Promise.* New York: E. P. Dutton.

Carlassare, Elizabeth (1994), 'Essentialism in Ecofeminist Discourse,' *Capitalism, Nature, Socialism,* 5 (3): 1–18.

Carruthers, David (2001), 'From Opposition to Orthodoxy: The Remaking of Sustainable Development,'

Journal of Third World Studies, 18 (2): 93–112.

Carson, Lyn, and Martin, Brian (1999), *Random Selection in Politics*. Westport, Conn.: Praegar.

Castells, Manuel (1996), *The Rise of the Network Society*. Oxford: Basil Blackwell.

Catton, William R. (1980), *Overshoot: The Ecological Basis of Revolutionary Change*. Urbana: University of Illinois Press.

Chock, Phyllis Pease (1995), 'Ambiguity in Policy Discourse: Congressional Talk About Immigration,' *Policy Sciences*, 28: 165–84.

Christ, Carol P. (1990), 'Rethinking Theology and Nature,' pp. 58–69 in Irene Diamond and Gloria Feman Orenstein (eds.), *Reweaving the World: The Emergence of Ecofeminism*. San Francisco, Calif.: Sierra Club Books.

Christoff, Peter (1995), 'Whatever Happened to Ecologically Sustainable Development?,' *Capucchino Papers* (Australian Conservation Foundation), 1: 69–74.

Christoff, Peter (1996*a*), 'Ecological Modernisation, Ecological Modernities,' *Environmental Politics*, 5: 476–500.

—— (1996*b*), 'Ecological Citizens and Ecologically Guided Democracy,' pp. 151–69 in Brian Doherty and Marius de Geus (eds.), *Democracy and Green Political Thought*. London: Routledge.

Coase, Ronald H. (1960), 'The Problem of Social Cost,' *Journal of Law and Economics*, 3: 1–44.

Cohen, Bernard L. (1984), 'Statement of Dissent', p. 556 in Julian L. Simon and Herman Kahn (eds.), *The Resourceful Earth: A Repsonse to Global 2000*. New York: Basil Blackwell.

Coleman, James S. (1986), 'Social Theory, Social Research, and a Theory of Action,' *American Journal of Sociology*, 91: 1309–35.

Commoner, Barry (1972), *The Closing Circle*. New York: Bantam.

Conca, Ken (1994), 'Peace, Justice, and Sustainability,' *Newsletter of the Committee on the Political Economy of the Good Society*, 4 (1): 1–7.

——, Princen, Thomas, and Maniates, Michael F. (2001), 'Confronting Consumption,' *Global Environmental Politics*, 1 (3): 1–10.

Crist, Eileen (2004), 'Against the Social Construction of Nature and Wilderness,' *Environmental Ethics*, 26: 5–24.

Cronon, William (1995), 'The Trouble with Wilderness; or, Getting Back to the Wrong Nature,' pp. 69–90 in William Cronon (ed.), *Uncommon Ground: Rethinking the Human Place in Nature*. New York: W.W. Norton.

Czech, Brian (2000), *Shoveling Fuel for a Runaway Train: Errant Economists, Shameful Spenders, and a Plan to Stop Them All*. Berkeley: University of California Press.

Daly, Herman E. (1977), *Steady-State Economics*. San Francisco: W. H. Freeman.

—— (1992), 'Free Market Environmentalism: Turning a Good Servant into a Bad Master,' *Critical Review*, 6: 171–83.

—— (1993), 'Sustainable Growth: An Impossibility Theorem,' pp. 267–73 in Herman E. Daly and Kenneth E. Townsend (eds.), *Valuing the Earth: Economics, Ecology, Ethics*. Cambridge, Mass.: MIT Press.

Daugbjerg, Carsten, and Svendsen, Gert Tinggaard (2003), 'Designing Green Taxes in a Political Context: From

Optimal to Feasible Environmental Regulation,' *Environmental Politics*, 12 (4): 76–95.
Davidson, Carlos (2000), 'Economic Growth and the Environment: Alternatives to the Limits Paradigm,' *Bioscience*, 50 (5): 433–8.
Davis, Karen (1995), 'Thinking Like a Chicken: Farm Animals and the Feminine Connection,' pp. 192–212 in Carol J. Adams and Josephine Donovan (eds.), *Animals and Women: Feminist Theoretical Explorations*. Durham, NC: Duke University Press.
Department of the Environment (United Kingdom) (1988), *Our Common Future: A Perspective by the United Kingdom on the Report of the World Commission on Environment and Development*. London: HMSO.
de-Shalit, Avner (2000), *The Environment Between Theory and Practice*. Oxford: Oxford University Press.
Devall, Bill, and Sessions, George (1985), *Deep Ecology: Living as if Nature Mattered*. Salt Lake City, Utah: Peregrine Smith.
Diamond, Irene (1994), *Fertile Ground: Women, Fertility, and the Living Earth*. Boston, Mass.: Beacon.
—— and Feman Orenstein, Gloria (eds.) (1990), *Reweaving the World: The Emergence of Ecofeminism*. San Francisco, Calif.: Sierra Club Books.
DiZerega, Gus (1993), 'Unexpected Harmonies: Self-Organization in Liberal Modernity and Ecology,' *The Trumpeter*, 10: 25–32.
Dobson, Andrew (1990), *Green Political Thought: An Introduction*. London: Unwin Hyman.
—— (1998), *Justice and the Environment: Conceptions of Environmental Sustainability and Social Justice*. Oxford: Oxford University Press.
—— (2004), *Citizenship and the Environment*. Oxford: Oxford University Press.
Dodge, Jim (1981), 'Living by Life: Some Bioregional Theory and Practice,' *Coevolution Quarterly*, 32: 6–12.
Doherty, Brian (1999), 'Paving the Way: The Rise of Direct Action Against Road-Building and the Changing Character of British Environmentalism,' *Political Studies*, 47: 275–91.
Dowie, Mark (1995), *Losing Ground: American Environmentalism at the Close of the Twentieth Century*. Cambridge, Mass.: MIT Press.
Dryzek, John S. (1987), *Rational Ecology: Environment and Political Economy*. New York: Basil Blackwell.
—— (1990*a*), 'Green Reason: Communicative Ethics for the Biosphere,' *Environmental Ethics*, 12: 195–210.
—— (1990*b*), *Discursive Democracy: Politics, Policy, and Political Science*. New York: Cambridge University Press.
—— (1992*a*), 'The Good Society versus the State: Freedom and Necessity in Political Innovation,' *Journal of Politics*, 54: 518–40.
—— (1992*b*), 'How far is it From Virginia and Rochester to Frankfurt? Public Choice as Critical Theory,' *British Journal of Political Science*, 22: 397–417.
—— (1992*c*), 'Ecology and Discursive Democracy: Beyond Liberal Capitalism and the Administrative State,' *Capitalism, Nature, Socialism*, 3 (2): 18–42.
—— (1995), 'The Informal Logic of Institutional Design,' pp. 103–25 in Robert E. Goodin (ed.), *The Theory*

of Institutional Design. New York: Cambridge University Press.

DRYZEK, JOHN S. (1996*a*), *Democracy in Capitalist Times: Ideals, Limits, and Struggles*. New York: Oxford University Press.

—— (1996*b*), 'Foundations for Environmental Political Economy: The Search for Homo Ecologicus?,' *New Political Economy*, 1: 27–40.

—— (1996*c*), 'Political and Ecological Communication,' pp. 13–30 in Freya Mathews (ed.), *Ecology and Democracy*. London: Frank Cass.

—— (1996*d*), 'Strategies of Ecological Democratization,' pp. 108–23 in William M. Lafferty and James Meadowcroft (eds.), *Democracy and the Environment: Problems and Prospects*. Cheltenham: Edward Elgar.

——, DOWNES, DAVID, HUNOLD, CHRISTIAN, and SCHLOSBERG, DAVID, with HERNES, HANS-KRISTIAN (2003), *Green States and Social Movements: Environmentalism in the United States, United Kingdom, Germany, and Norway*. Oxford: Oxford University Press.

EASTERBROOK, GREG (1995), *A Moment on the Earth: The Coming Age of Environmental Optimism*. New York: Penguin.

ECKERSLEY, ROBYN (1992), *Environmentalism and Political Theory: Toward an Ecocentric Approach*. Albany: State University of New York Press.

—— (ed.) (1995), *Markets, the State and the Environment: Towards Integration*. Melbourne: Macmillan.

—— (2004), *The Green State: Rethinking Democracy and Sovereignty*. Cambridge, Mass.: MIT Press.

EHRLICH, PAUL (1968), *The Population Bomb*. New York: Ballantine.

—— and EHRLICH, ANNE H. (1974), *The End of Affluence: A Blueprint for Your Future*. New York: Ballantine.

—— (1981), *Extinction: The Causes and Consequences of the Disappearance of Species*. New York: Random House.

—— (1996), *The Betrayal of Science and Reason*. Washington, DC: Island Press.

—— (2004), *One with Nineveh: Politics, Consumption, and the Human Future*. Washington, DC: Island Press.

ENVIRONMENT 2004 (2003), *The Bush Environmental Record: An Unprecedented Assault on America's Health and Heritage*. Washington, DC: Environment 2004.

EUROPEAN ENVIRONMENT AGENCY (2000), *Environmental Taxation: Recent Developments in Tools for Integration*. Copenhagen: European Environment Agency.

FIORINO, DANIEL J. (1995), *Making Environmental Policy*. Berkeley, Calif.: University of California Press.

—— (2004), 'Flexibility,' pp. 393–425 in Robert F. Durant, Daniel J. Fiorino, and Rosemary O'Leary (eds.), *Environmental Governance Reconsidered*. Cambridge, Mass.: MIT Press.

FISCHER, FRANK (1993), 'Citizen Participation and the Democratization of Policy Expertise: From Theoretical Inquiry to Practical Cases,' *Policy Sciences*, 26: 165–87.

FISCHOFF, BARUCH, SLOVIC, PAUL, and LICHTENSTEIN, SARAH (1982), 'Lay Fables and Expert Foibles in Judgments About Risk,' *American Statistician*, 36: 240–55.

FOREMAN, DAVE (1985), *Ecodefense: A Field Guide to Monkeywrenching*. Tucson, Ariz.: Ned Ludd Books.

—— (1991), Contribution to Murray Bookchin and Dave Foreman, *Defending the Earth: A Dialogue*

Between Murray Bookchin and Dave Foreman. Boston, Mass.: South End Press.

—— (1998), 'Wilderness Areas for Real,' pp. 395–407 in J. Baird Callicott and Michael P. Nelson (eds.), *The Great New Wilderness Debate*. Athens, GA: University of Georgia Press.

—— (2000), 'The Real Wilderness Idea,' *USDA Forest Service Proceedings*, RMRS-P-15-Vol-1.

FOUCAULT, MICHEL (1980), *Power/Knowledge: Selected Interviews and Other Writings, 1972–1977*. Brighton: Harvester.

FOX, WARWICK (1984), 'On Guiding Stars to Deep Ecology: A Reply to Naess,' *The Ecologist*, 14: 203–4.

—— (1990), *Toward a Transpersonal Ecology: Developing New Foundations for Environmentalism*. Boston, Mass.: Shambhala.

FREEMAN, CHRISTOPHER (1973), 'Malthus with a Computer,' in H. S. D. Cole, C. Freeman, M. Jahoda, and K. L. R. Pravitts (eds.), *Models of Doom: A Critique of The Limits to Growth*. New York: Universe.

GARNER, ROBERT (1993), *Animals, Politics and Morality*. Manchester: Manchester University Press.

GEORGESCU-ROEGEN, NICHOLAS (1971), *The Entropy Law and the Economic Process*. Cambridge, Mass.: Harvard University Press.

GERTH, H. H., and WRIGHT MILLS, C. (1948), *From Max Weber: Essays in Sociology*. London: Routledge and Kegan Paul.

GONZÁLEZ, GEORGE A. (2001a), *Corporate Power and the Environment: The Political Economy of US Environmental Policy*. Lanham, Md.: Rowman and Littlefield.

—— (2001b), 'Democratic Ethics and Ecological Modernization: The Formulation of California's Automobile Emission Standards,' *Public Integrity*, 3: 325–44.

—— (2002), 'Local Growth Coalitions and Air Pollution Controls: The Ecological Modernisation of the US in Historical Perspective,' *Environmental Politics*, 11 (3): 121–44.

GOODIN, ROBERT E. (1992), *Green Political Theory*. Cambridge: Polity.

—— (1994), 'Selling Environmental Indulgences,' *Kyklos*, 47: 573–95.

GORDON, H. SCOTT (1954), 'The Economic Theory of a Common-Property Resource: The Fishery,' *Journal of Political Economy*, 62: 124–42.

GORE, ALBERT (1992), *Earth in the Balance*. Boston, Mass.: Houghton Mifflin.

GOTTLEIB, ROGER S. (ed.) (1996), *This Sacred Earth: Religion, Nature, Environment*. New York: Routledge.

GRAY, TIM (2000), 'A Discourse Analysis of UK Contaminated Land Policy,' paper presented at the Conference of the Political Studies Association, London, April 10–13.

GREGG, ALAN (1955), 'A Medical Aspect of the Population Problem,' *Science*, 121: 681–2.

GUHA, RAMACHANDRA (1997), 'The Environmentalism of the Poor,' pp. 3–21 in Ramachandra Guha and Juan Martinez-Alier (eds.), *Varieties of Environmentalism*. London: Earthscan.

GUNDERSEN, ADOLF (1995), *The Environmental Promise of Democratic Deliberation*. Madison: University of Wisconsin Press.

HAAS, PETER M. (1992), 'Banning Chlorofluorocarbons: Efforts to Protect Stratospheric Ozone,' *International Organization*, 46: 187–224.

Haas, Peter M., Keohane, Robert O., and Levy, Marc A. (1993), *Institutions for the Earth: Sources of Effective Environmental Protection*. Cambridge, Mass.: MIT Press.

Hahn, Robert W. (1995), 'Economic Prescriptions for Environmental Policy Instruments: Lessons from the United States and Continental Europe,' pp. 129–56 in Robyn Eckersley (ed.), *Markets, the State, and the Environment: Towards Integration*. Melbourne: Macmillan.

Hajer, Maarten A. (1995), *The Politics of Environmental Discourse: Ecological Modernization and the Policy Process*. Oxford: Oxford University Press.

—— (2003), 'A Frame in the Fields,' pp. 88–110 in Marten A. Hajer and Hendrik Wagenaar (eds.), *Deliberative Policy Analysis: Understanding Governance in the Network Society*. Cambridge: Cambridge University Press.

Hardin, Garrett (1968), 'The Tragedy of the Commons,' *Science*, 162: 1243–8.

—— (1977), 'Living on a Lifeboat,' pp. 261–79 in Garrett Hardin and John Baden (eds.), *Managing the Commons*. San Francisco, Calif.: W. H. Freeman.

—— (1993), *Living Within Limits: Ecology, Economics, and Population Taboos*. New York: Oxford University Press.

Hawken, Paul, Lovins, Amory B., and Lovins, L. Hunter (1999), *Natural Capitalism: Creating the Next Industrial Revolution*. Boston, Mass.: Little, Brown.

Hay, Peter (2002), *Main Currents in Western Environmental Thought*. Sydney: University of New South Wales Press.

Hayek, Friedrich A. von (1979), *Law, Legislation, and Liberty: The Political Order of a Free People*. Chicago: University of Chicago Press.

Haynes, Jeff (1999), 'Power, Politics and Environmental Movements in the Third World,' *Environmental Politics*, 8 (1): 222–42.

Hays, Samuel P. (1959), *Conservation and The Gospel of Efficiency: The Progressive Conservation Movement, 1890–1920*. Cambridge, Mass.: Harvard University Press.

—— (1987), *Beauty, Health, and Permanence: Environmental Politics in the United States, 1955–1985*. Cambridge: Cambridge University Press.

Hayward, Tim (1995), *Ecological Thought: An Introduction*. Cambridge: Polity.

Heilbroner, Robert (1991), *An Inquiry into the Human Prospect: Looked at Again for the 1990s*. New York: Norton.

Holliday, Charles O., Jr., Schmidheiny, Stephan, and Watts, Philip (2002), *Walking the Talk: The Business Case for Sustainable Development*. Sheffield: Greenleaf.

Homer-Dixon, Thomas F. (1999), *Environment, Scarcity and Violence*. Princeton: Princeton University Press.

—— (2000), *The Ingenuity Gap: Can We Solve the Problems of the Future?* Toronto: Knopf.

Huber, Joseph (1982), *Die verlorene Unschuld der Okologie*. Frankfurt am Main: Fischer Verlag.

Inglehart, Ronald (1990), *Culture Shift in Advanced Industrial Society*. Princeton, NJ: Princeton University Press.

Innes, Judith E., and Booher, David E. (2003), 'Collaborative Policymaking: Governance Through Dialogue,' pp. 33–59 in Marten A. Hajer and Hendrik Wagenaar (eds.), *Deliberative Policy Analysis: Understanding Governance in the Network Society*. Cambridge: Cambridge University Press.

INTERNATIONAL UNION FOR THE CONSERVATION OF NATURE (1980), *World Conservation Strategy*. Gland, Switzerland: International Union for the Conservation of Nature.

JACOBS, MICHAEL (1991), *The Green Economy: Environment, Sustainable Development and the Politics of the Future*. London: Pluto.

—— (1995), 'Financial Incentives: The British Experience,' pp. 113–28 in Robyn Eckersley (ed.), *Markets, the State, and the Environment: Towards Integration*. Melbourne: Macmillan.

—— (1999), *Environmental Modernisation: The New Labour Agenda*. London: Fabian Society.

JAHN, DETLEF (1998), 'Environmental Performance and Policy Regimes: Explaining Variations in 18 OECD Countries,' *Policy Sciences*, 31: 107–31.

JÄNICKE, MARTIN (1985), *Preventive Environmental Policy as Ecological Modernization and Structural Policy*. Berlin: Wissenschaftszentrum.

—— (1992), 'Conditions for Environmental Policy Success: An International Comparison,' *The Environmentalist*, 12: 47–58.

—— (1996), 'Democracy as a Condition for Environmental Policy Success: The Importance of Non-institutional Factors,' pp. 71–85 in William M. Lafferty and James Meadowcroft (eds.), *Democracy and the Environment: Problems and Prospects*. Cheltenham: Edward Elgar.

—— and WEIDNER, HELMUT (1997), 'Germany,' pp. 133–55 in Martin Jänicke and Helmut Weidner (eds.), *National Environmental Policies: A Comparative Study of Capacity Building*. Berlin: Springer.

JEVONS, W. STANLEY (1865), *The Coal Question*. London: Macmillan.

JORDAN, ANDREW, WURZEL, RÜDIGER K. W., and ZITO, ANTHONY R. (2003), ' "New" Instruments of Environmental Governance: Patterns and Pathways of Change,' *Environmental Politics*, 12 (1): 1–24.

KELMAN, STEVEN (1981), *What Price Incentives? Economists and the Environment*. Boston, Mass.: Auburn House.

—— (1987), *Making Public Policy: A Hopeful View of American Government*. New York: Basic Books.

KEMP, RAY (1985), 'Planning, Public Hearings, and the Politics of Discourse,' pp. 177–201 in John Forester (ed.), *Critical Theory and Public Life*. Cambridge, Mass.: MIT Press.

KHEEL, MARTI (1990), 'Ecofeminism and Deep Ecology: Reflections on Identity and Difference,' pp. 128–37 in Irene Diamond and Gloria Feman Orenstein (eds.), *Reweaving the World: The Emergence of Ecofeminism*. San Francisco, Calif.: Sierra Club Books.

KNEESE, ALLEN V., and SCHULTZE, CHARLES L. (1975), *Pollution, Prices, and Public Policy*. Washington, DC: Brookings.

KOVEL, JOEL (2002), *The Enemy of Nature: The End of Capitalism or the End of the World*. London: Zed Books.

—— and LOWY, MICHAEL (2002), 'An Ecosocialist Manifesto,' *Capitalism, Nature, Socialism*, 31 (1): 1–2 and 155–7.

LA CHAPELLE, DOLORES (1978), *Earth Wisdom*. San Diego, Calif.: Guild of Tudors.

LAFFERTY, WILLIAM M. (1996), 'The Politics of Sustainable Development,' *Environmental Politics*, 5 (2): 185–208.

—— and HOVDEN, EIVIND (2003), 'Environmental Policy Integration: Towards an Analytical Framework,' *Environmental Politics*, 12 (3): 1–22.

LAFFERTY, WILLIAM M., and MEADOWCROFT, JAMES (eds.) (2000), *Implementing Sustainable Development: Strategies and Initiatives in High Consumption Societies*. Oxford: Oxford University Press.

LANGHELLE, OLUF (2000), 'Why Ecological Modernization and Sustainable Development Should Not be Conflated,' *Journal of Environmental Policy and Planning*, 2: 303–22.

LAPPÉ, FRANCES MOORE, and COLLINS, JOSEPH (1977), *Food First: Beyond the Myth of Scarcity*. Boston: Houghton Mifflin.

LEE, KAI N. (1993), *Compass and Gyroscope: Integrating Science and Politics for the Environment*. Washington, DC: Island Press.

—— (1999), 'Appraising Adaptive Management,' *Conservation Ecology*, 3 (2), online at **www.consecol.org/vol3/iss2/art3/**

LEHMBRUCH, GERHARD (1984), 'Concertation and the Structure of Corporatist Networks,' in John H. Goldthorpe (ed.), *Order and Conflict in Contemporary Capitalism*. Oxford: Clarendon.

LEWIS, MARTIN W. (1992), *Green Delusions: An Environmentalist Critique of Radical Environmentalism*. Durham, NC: Duke University Press.

LIGHT, ANDREW, and KATZ, ERIC (1996), *Environmental Pragmatism*. London: Routledge.

LINDBLOM, CHARLES E. (1959), 'The Science of Muddling Through,' *Public Administration Review*, 19: 79–88.

—— (1965), *The Intelligence of Democracy: Decision Making through Mutual Adjustment*. New York: Free Press.

—— (1977), *Politics and Markets: The World's Political-Economic Systems*. New York: Basic Books.

—— (1982), 'The Market as Prison,' *Journal of Politics*, 44: 324–36.

LITFIN, KAREN T. (1994), *Ozone Discourses: Science and Politics in Global Environmental Cooperation*. New York: Columbia University Press.

LOMBORG, BJØRN (2001*a*), *The Skeptical Environmentalist: Measuring the Real State of the World*. Cambridge: Cambridge University Press.

—— (2001*b*), 'The Truth About the Environment,' *The Economist*, 360 (August 4): 63.

LOVELOCK, JAMES E. (1979), *Gaia: A New Look at Life on Earth*. Oxford: Oxford University Press.

LUDWIG, DONALD, HILBORN, RAY, and WALTERS, CARL (1993), 'Uncertainty, Resource Expolitation, and Conservation: Lessons from History,' *Science*, 260.

LUKE, TIMOTHY (1997), *Ecocritique: Contesting the Politics of Nature, Economy, and Culture*. Minneapolis: University of Minnesota Press.

—— (1999), 'Environmentality as Green Governmentality,' pp. 121–51 in Eric Darier (ed.), *Discourses of the Environment*. Oxford: Basil Blackwell.

LUNDQVIST, LENNART J. (2004), *Sweden and Ecological Governance: Straddling the Fence*. Manchester: Manchester University Press.

MCFARLAND, ANDREW (1984), 'An Experiment in Regulatory Negotiation: The National Coal Policy Project,' paper presented at the Annual Meeting of the Western Political Science Association.

MCGINNIS, MICHAEL VONCENT (ed.) (1998), *Bioregionalism*. New York: Routledge.

MCPHEE, JOHN (1970), *Encounters with the Archdruid*. New York: Farrar, Straus, and Giroux.

MANES, CHRISTOPHER (1990), *Green Rage: Radical Environmentalism and the Unmaking of Civilization.* Boston, Mass.: Little, Brown.

MANIATES, MICHAEL F. (2001), 'Individualization: Plant a Tree, Buy a Bike, Save the World,' *Global Environmental Politics,* 1 (3): 31–52.

MASON, MICHAEL (1999), *Ecological Democracy.* London: Earthscan.

MEADOWCROFT, JAMES (2000), 'Sustainable Development: A New(ish) Idea for a New Century?' *Political Studies,* 48: 370–87.

MEADOWS, DONELLA H. (1976), 'Equity, The Free Market, and the Sustainable State,' in Dennis L. Meadows (ed.), *Alternatives to Growth, I: Toward a Sustainable Future.* Cambridge, Mass.: Ballinger.

——, MEADOWS, DENNIS L., RANDERS, JORGEN, and BEHRENS, WILLIAM H. III (1972), *The Limits to Growth.* New York: Universe Books.

——, MEADOWS, DENNIS L., and RANDERS, JORGEN (1992), *Beyond the Limits: Confronting Global Collapse, Envisioning a Sustainable Future.* Post Mills, Vt.: Chelsea Green.

MEIDINGER, ERROL E. (2003), 'Forest Certification as a Global Civil Society Regulatory Institution,' pp. 265–89 in Errol E. Meidinger, Chris Elliott, and Gerhard Oesten (eds.), *Social and Political Dimensions of Forest Certification.* Remagen-Oberwinter: Forstbuch Verlag.

MEINERS, ROGER E., and YANDLE, BRUCE (1993), 'Taking the Environment Seriously: What Do We Mean?,' pp. vii–xiv in Roger E. Meiners and Bruce Yandle (eds.), *Taking the Environment Seriously.* Lanham, Md.: Rowman and Littlefield.

MERCHANT, CAROLYN (1992), *Radical Ecology.* London: Routledge.

MICHAELS, PATRICK J. (1993), 'Global Warming: Facts vs. The Popular Vision,' pp. 341–62 in David Boaz and Edward H. Crane (eds.), *Market Liberalism: A Paradigm for the 21st Century.* Washington, DC: Cato Institute.

MILBRATH, LESTER W. (1989), *Envisioning a Sustainable Society: Learning Our Way Out.* Albany: State University of New York Press.

MITCHELL, WILLIAM C., and SIMMONS, RANDY T. (1994), *Beyond Politics: Markets, Welfare, and the Failure of Bureaucracy.* Boulder, Colo.: Westview.

MOL, ARTHUR P. J. (1996), 'Ecological Modernisation and Institutional Reflexivity: Environmental Reform in the Late Modern Age,' *Environmental Politics,* 5: 302–23.

—— and SPAARGAREN, GERT (2000), 'Ecological Modernisation Theory in Debate: A Review,' *Environmental Politics,* 9 (1): 17–49.

MORAN, ALAN (1995), 'Tools for Environmental Policy: Market Instruments versus Command-and-Control,' pp. 73–85 in Robyn Eckersley (ed.), *Markets, the State, and the Environment: Towards Integration.* Melbourne: Macmillan.

MOSHER, LAWRENCE (1983), 'Distrust of Gorsuch May Stymie EPA Attempt to Integrate Pollution Wars,' *National Journal,* 15: 322–4.

MURPHY, E. F. (1967), *Governing Nature.* Chicago, Ill.: Quadrangle Books.

MYERS, NORMAN, and SIMON, JULIAN L. (1994), *Scarcity or Abundance: A Debate on the Environment.* New York: Norton.

NAESS, ARNE (1973), 'The Shallow and the Deep, Long-Range Ecology Movement: A Summary,' *Inquiry,* 16: 95–100.

NAESS, ARNE (1989), *Ecology, Community and Lifestyle*. Cambridge: Cambridge University Press.

NEF, JOHN U. (1977), 'An Early Energy Crisis and its Consequences,' *Scientific American*, 237: 140–51.

NELKIN, DOROTHY, and POLLACK, MICHAEL (1981), *The Atom Besieged*. Cambridge, Mass.: MIT Pess.

NELSON, ROBERT H. (1993), 'How Much is Enough? An Overview of the Benefits and Costs of Environmental Protection,' pp. 1–23 in Roger E. Meiners and Bruce Yandle (eds.), *Taking the Environment Seriously*. Lanham, Md.: Rowman and Littlefield.

O'BRIEN, MARY (2000), *Making Better Environmental Decisions: An Alternative to Risk Assessment*. Cambridge, Mass.: MIT Press.

O'CONNOR, JAMES (1988), 'Capitalism, Nature, Socialism: A Theoretical Introduction,' *Capitalism, Nature, Socialism*, 1: 11–38.

OFFE, CLAUS (1985), 'New Social Movements: Challenging the Boundaries of Institutional Politics,' *Social Research*, 52: 817–68

—— (1990), 'Reflections on the Institutional Self-Transformation of Movement Politics: A Tentative Stage Model,' pp. 232–50 in Russell J. Dalton and Manfred Kuechler (eds.), *Challenging the Political Order: New Social and Political Movements in Western Democracies*. New York: Oxford University Press.

OLSON, MANCUR (1965), *The Logic of Collective Action*. Cambridge, Mass.: Harvard University Press.

OPHULS, WILLIAM (1977), *Ecology and the Politics of Scarcity*. San Francisco, Calif.: W. H. Freeman.

—— and STEPHEN BOYAN, A., JR. (1992), *Ecology and the Politics of Scarcity Revisited*. San Francisco, Calif.: W. H. Freeman.

OPSCHOOR, J. B., and VOS, H. (1988), *The Application of Economic Instruments for Environmental Protection in OECD Member Countries*. Paris: OECD.

ORGANIZATION FOR ECONOMIC COOPERATION AND DEVELOPMENT (1989), *Economic Instruments for Environmental Protection*. Paris: OECD.

OSTROM, ELINOR (1990), *Governing the Commons*. Cambridge: Cambridge University Press.

PAEHLKE, ROBERT (1988), 'Democracy, Bureaucracy, and Environmentalism,' *Environmental Ethics*, 10: 291–308.

PATERSON, MATTHEW (2001), 'Risky Business: Insurance Companies in Global Warming,' *Global Environmental Politics*, 1 (4): 18–42.

PEARCE, DAVID, and BARBIER, EDWARD R. (2000), *Blueprint for a Sustainable Economy*. London: Earthscan.

PEARCE, DAVID, MARKANDYA, ANIL, and BARBIER, EDWARD R. (1989), *Blueprint for a Green Economy*. London: Earthscan.

PECCEI, AURELIO (1981), *One Hundred Pages for the Future*. New York: Mentor.

PEPPER, DAVID (1993), *Ecosocialism: From Deep Ecology to Social Justice*. London: Routledge.

—— (1999), 'Ecological Modernisation or the "Ideal Model" of Sustainable Development? Questions Prompted at Europe's Periphery,' *Environmental Politics*, 8 (4): 1–34.

PICKEL, ANDREAS (1993), 'Authoritarianism or Democracy? Marketization as a Political Problem,' *Policy Sciences*, 26: 139–63.

PLANT, JUDITH (ed.) (1989), *Healing the Wounds: The Promise of Ecofeminism*. Philadelphia, Pa.: New Society Publishers.

PLUMWOOD, VAL (1993), *Feminism and the Mastery of Nature*. London: Routledge.

—— (1995), 'Has Democracy Failed Ecology? An Ecofeminist Perspective,' *Environmental Politics*, 4 (4): 134–68.

—— (2002), *Environmental Culture: The Ecological Crisis of Reason*. London: Routledge.

POGUNTKE, THOMAS (2002), 'Green Parties in National Governments: From Protest to Acquiescence,' *Environmental Politics*, 11 (1): 133–45.

POPPER, KARL R. (1966), *The Open Society and its Enemies*. London: Routledge and Kegan Paul.

—— (1972), *The Poverty of Historicism*, revised edition. London: Routledge and Kegan Paul.

PORRITT, JONATHAN (1986), *Seeing Green: The Politics of Ecology Explained*. Oxford: Basil Blackwell.

PRESS, DANIEL (1994), *Democratic Dilemmas in the Age of Ecology: Trees and Toxics in the American West*. Durham, NC: Duke University Press.

PRESSMAN, JEFFREY, and WILDAVSKY, AARON (1973), *Implementation*. Berkeley, Calif.: University of California Press.

RABE, BARRY G. (1991), 'Beyond the Nimby Syndrome in Hazardous Waste Facility Siting: The Albertan Breakthrough and the Prospects for Cooperation in Canada and the United States,' *Governance*, 4: 184–206.

REGAN, TOM (1983), *The Case for Animal Rights*. Berkeley, Calif.: University of California Press.

REISNER, MARC (1993), *Cadillac Desert: The American West and its Disappearing Water*, revised edition. New York: Penguin.

REUVENY, RAFAEL (2002), 'Economic Growth, Environmental Scarcity, and Conflict,' *Global Environmental Politics*, 2 (1): 83–110.

REVELL, ANDREA (2003), 'Is Japan an Ecological Frontrunner Nation?' *Environmental Politics*, 12 (4): 24–48.

RICHARDSON, DICK (1994), 'The Politics of Sustainable Development,' paper presented to the International Conference on the Politics of Sustainable Development within the European Union, University of Crete, Greece.

RIDLEY, MATT (1995), *Down to Earth: A Contrarian View of Environmental Problems*. London: Institute of Economic Affairs.

ROODMAN, DAVID MALIN (1996), 'Harnessing the Market for the Environment,' pp. 168–87 in Lester R. Brown (ed.), *State of the World 1996*. New York: W. W. Norton.

ROOTES, CHRISTOPHER (2003), 'The Transformation of Environmental Activism: An Introduction,' pp. 1–11 in Christopher Rootes (ed.), *Environmental Protest in Western Europe*. Oxford: Oxford University Press.

ROSENBAUM, WALTER A. (1985), *Environmental Politics and Policy*. Washington, DC: Congressional Quarterly Press.

—— (1995), 'The Bureaucracy and Environmental Policy,' pp. 206–41 in James P. Lester (ed.), *Environmental Politics and Policy: Theories and Evidence*, 2nd edn. Durham, NC: Duke University Press.

ROWELL, ANDREW (1996), *Green Backlash: Global Subversion of the Environmental Movement*. London: Routledge.

RUSSELL, JULIA SCOFIELD (1990), 'The Evolution of an Ecofeminist,' pp. 223–30 in Irene Diamond and Gloria Feman

Orenstein (eds.), *Reweaving the World: The Emergence of Ecofeminism.* San Francisco, Calif.: Sierra Club Books.

RYLE, MARTIN (1988), *Ecology and Socialism.* London: Century Hutchinson.

SABEL, CHARLES, FUNG, ARCHON, and KARKKAINEN, BRADLEY (2000), 'Beyond backyard Environmentalism: How Communities are Quietly Refashioning Environmental Regulation,' *Boston Review,* online at **www.bostonreview.mit.edu/BR24.5/sabel.html**

SACHS, WOLFGANG (1999), 'Sustainable Development and the Crisis of Nature: On the Political Anatomy of an Oxymoron,' pp. 23–41 in Frank Fischer and Maarten A. Hajer (eds.), *Living with Nature: Environmental Politics as Cultural Discourse.* Oxford: Oxford University Press.

SAGOFF, MARK (1988), *The Economy of the Earth.* Cambridge: Cambridge University Press.

SAIRINEN, RAUNO (2003), 'The Politics of Regulatory Reform: "New" Environmental Policy Instruments in Finland,' *Environmental Politics,* 12 (4): 24–48.

SALE, KIRKPATRICK (1985), *Dwellers in the Land: The Bioregional Vision.* San Francisco, Calif.: Sierra Club Books.

SANDEL, MICHAEL J. (1997), 'It's Immoral to Buy the Right to Pollute,' *New York Times,* December 15: A29.

SANDILANDS, CATRIONA (1999), 'Sex at the Limits,' pp. 79–94 in Eric Darier (ed.), *Discourses of the Environment.* Oxford: Basil Blackwell.

SCHEINBERG, ANNE (2003), 'The Proof of the Pudding: Urban Recycling in North America as a Process of Ecological Modernisation,' *Environmental Politics,* 12 (4): 49–75.

SCHLOSBERG, DAVID (1999), *Environmental Justice and the New Pluralism: The Challenge of Difference for Environmentalism.* Oxford: Oxford University Press.

SCHMIDHEINY, STEPHAN (1992), *Changing Course: A Global Business Perspective on Development and the Environment.* Cambridge, Mass.: MIT Press.

SCHMIDT, CHARLES W. (2003), 'Subjective Science: Environmental Cost-Benefit Analysis,' *Environmental Health Perspectives,* 110 (10): 530–2.

SCHMITT, CARL (1986), *Political Romanticism.* Cambridge, Mass.: MIT Press.

SCHUMACHER, E. F. (1973), *Small is Beautiful: Economics as if People Mattered.* New York: Harper and Row.

SCHWARTZ, PETER, and RANDALL, DOUG (2003), *An Abrupt Climate Change Scenario and its Implications for National Security.* Washington, DC: United States Department of Defense. Online at **www.ems.org/climate/pentagon_climatechange.pdf**

SCRUGGS, LYLE (2001), 'Is There Really a Link Between Neo-Corporatism and Environmental Performance? Updated Evidence and New Data for the 1980s and 1990s,' *British Journal of Political Science,* 31: 686–92.

SHIVA, VANDANA (2000). *Poverty and Globalisation.* BBC Reith Lecture, online at **www.news.bbc.co.uk/hi/english/static/events/reith_2000/lecture5.stm**.

SIMON, HERBERT A. (1981), *The Sciences of the Artificial,* 2nd edn. Cambridge, Mass.: MIT Press.

SIMON, JULIAN (1981), *The Ultimate Resource.* Princeton, NJ: Princeton University Press.

—— (1996), *The Ultimate Resource 2.* Princeton, NJ: Princeton University Press.

—— and Kahn, Herman (eds.) (1984), *The Resourceful Earth: A Response to Global 2000.* New York: Basil Blackwell.

Singer, Peter (1975), *Animal Liberation.* New York: Avon.

Smith, Graham (2003), *Deliberative Democracy and the Environment.* London: Routledge.

Smith, V. Kerry (ed.) (1979), *Scarcity and Growth Reconsidered.* Baltimore, Md.: Johns Hopkins University Press for Resources for the Future.

Soper, Kate (1995), *What is Nature? Culture, Politics, and the Non-Human.* Oxford: Basil Blackwell.

Spretnack, Charlene (1986), *The Spiritual Dimension of Green Politics.* Santa Fe, N. Mex.: Bear and Co.

Starhawk (1987), *Truth or Dare: Encounters with Power, Authority and Mystery.* San Francisco, Calif.: Harper and Row.

Stavins, Robert N. (2002). 'Lessons from the American Experience with Market-Based Environmental Instruments,' in John D. Donahue and Joseph S. Nye, Jr. (eds.), *Market-Based Governance.* Washington, DC: Brookings.

—— (2003) 'Taking Fish to Market,' *Forbes*, April 28.

Stretton, Hugh (1976), *Capitalism, Socialism, and the Environment.* Cambridge: Cambridge University Press.

Stroup, Richard L., and Meiners, Roger E. (2000), *Cutting Green Tape: Toxic Pollutants, Environmental Regulation, and the Law.* Oakland, Calif.: The Independent Institute.

Stroup, Richard L., and Shaw, Jane S. (1993), 'Environmental Harms from Federal Government Policy,' pp. 51–72 in Roger E. Meiners and Bruce Yandle (eds.), *Taking the Environment Seriously.* Lanham, Md.: Rowman and Littlefield.

Sugden, Robert, and Williams, Alan (1978), *The Principles of Practical Cost-Benefit Analysis.* Oxford: Oxford University Press.

Susskind, Lawrence, Levy, Paul F., and Thomas, Jennifer (2000), *Negotiating Environmental Agreements: How to Avoid Escalating Confrontation, Needless Costs, and Unnecessary Litigation.* Washington, DC: Island Press.

Szasz, Andrew (1994), *Ecopopulism: Toxic Waste and the Movement for Environmental Justice.* Minneapolis, Minn.: University of Minnesota Press.

Taylor, Bob Pepperman (1992), *Our Limits Transgressed.* Lawrence, Kan.: University Press of Kansas.

Taylor, Jerry (1993), 'The Growing Abundance of Natural Resources,' pp. 363–78 in David Boaz and Edward H. Crane (eds.), *Market Liberalism: A Paradigm for the 21st Century.* Washington, DC: Cato Institute.

Tesh, Sylvia N. (1993), 'New Social Movements and New Ideas,' paper presented at the Annual Meeting of the American Political Science Association, Washington, DC, September 2–5.

—— (2000), *Uncertain Hazards: Environmental Activists and Scientific Proof.* Ithaca, NY: Cornell University Press.

Thiele, Leslie Paul (1999), *Environmentalism for a New Millennium: The Challenge of Coevolution.* New York: Oxford University Press.

Thompson, Michael (1993), 'The Meaning of Sustainable Development,' paper presented to the Conference on

Global Governability, London School of Economics.

TORGERSON, DOUGLAS (1990), 'Limits of the Administrative Mind: The Problem of Defining Environmental Problems,' pp. 115–61 in Robert Paehlke and Douglas Torgerson (eds.), *Managing Leviathan: Environmental Politics and the Administrative State.* Peterborough, Ontario: Broadview.

—— (1994), 'Strategy and Ideology in Environmentalism: A Decentered Approach to Sustainability,' *Industrial and Environmental Crisis Quarterly*, 8: 295–321.

—— (1995), 'The Uncertain Quest for Sustainability: Public Discourse and the Politics of Environmentalism,' pp. 3–20 in Frank Fischer and Michael Black (eds.), *Greening Environmental Policy: The Politics of a Sustainable Future.* Liverpool: Paul Chapman.

—— (1999), *The Promise of Green Politics: Environmentalism and the Public Sphere.* Durham, NC: Duke University Press.

—— (2003), 'Democracy Through Policy Discourse,' pp. 113–38 in Marten A. Hajer and Hendrik Wagenaar (eds.), *Deliberative Policy Analysis: Understanding Governance in the Network Society.* Cambridge: Cambridge University Press.

—— and PAEHLKE, ROBERT (1990), 'Environmental Administration: Revising the Agenda of Inquiry and Practice,' pp. 7–16 in Robert Paehlke and Douglas Torgerson (eds.), *Managing Leviathan: Environmental Politics and the Administrative State.* Peterborough, Ontario: Broadview.

VAN MUIJEN, MARIE-LOUISE (2000), 'The Netherlands: Ambitious on Goals—Ambivalent on Action,' pp. 142–73 in William M. Lafferty and James Meadowcroft (eds.), *Implementing Sustainable Development: Strategies and Initiatives in High Consumption Societies.* Oxford: Oxford University Press.

VIG, NORMAN J., and KRAFT, MICHAEL E. (1984), *Environmental Policy in the 1980s: Reagan's New Agenda.* Washington, DC: Congressional Quarterly Press.

VOGEL, DAVID (1986), *National Styles of Regulation: Environmental Policy in Great Britain and the United States.* Ithaca, NY: Cornell University Press.

—— (2003), 'Comparing Environmental Governance: Risk Regulation in the EU and US,' paper presented at the Conference on Environmental Policy Integration and Sustainable Development, National Europe Centre, Australian National University, November 19–20.

VON FRANTZIUS, INA (2004), 'World Summit on Sustainable Development Johannesburg 2002: A Critical Assessment of the Outcomes,' *Environmental Politics*, 13: 467–73.

VON WEIZSÄCKER, ERNST, LOVINS, AMORY B., and LOVINS, L. HUNTER (1997), *Factor Four: Doubling Wealth—Halving Resource Use.* London: Earthscan.

WACKERNAGEL, MATHIS et al. (2002), 'Tracking the Ecological Overshoot of the Human Economy,' *Proceedings of the National Academy of Sciences*, 99 (14): 9266–71.

WAPNER, PAUL (1996), *Environmental Activism and World Civic Politics.* Albany: State University of New York Press.

—— (2002), 'Horizontal Politics: Transnational Environmental Activism and Global Cultural Change,' *Global Environmental Politics*, 2 (2): 37–62.

—— (2003), 'World Summit on Sustainable development: Toward a Post-Jo'burg Environmentalism,' *Global Environmental Politics*, 3 (1): 1–10.

Weale, Albert (1992), *The New Politics of Pollution*. Manchester: Manchester University Press.

—— (2001), 'Can we Democratise Decisions on Risk and the Environment? *Government and Opposition*, 36: 355–78.

White, Lynn, Jr. (1967), 'The Historical Roots of our Ecologic Crisis,' *Science*, 155: 1203–7.

White, Rob (1994), 'Green Politics and the Question of Population,' *Journal of Australian Studies*, 40: 27–43.

Wiesenthal, Herbert (1993), *Realism in Green Politics: Social Movements and Ecological Reform in Germany*. New York: St. Martin's.

Wildavsky, Aaron (1988), *The New Politics of the Budgetary Process*. Boston, Mass.: Little, Brown.

—— (1995), *But Is It True? A Citizen's Guide to Environmental Health and Safety Issues*. Cambridge, Mass.: Harvard University Press.

Williams, Bruce A., and Matheny, Albert R. (1995), *Democracy, Dialogue, and Environmental Disputes: The Contested Languages of Social Regulation*. New Haven, Conn.: Yale University Press.

Wissenburg, Marcel (1998), *Green Liberalism: The Free and the Green Society*. London: UCL Press.

World Bank (1995), *Monitoring Environmental Progress*. Washington, DC: The World Bank.

World Commission on Environment and Development (1987), *Our Common Future*. Oxford: Oxford University Press.

Yandle, Bruce (1993), 'Community Markets to Control Nonpoint Source Pollution,' pp. 185–207 in Roger E. Meiners and Bruce Yandle (eds.), *Taking the Environment Seriously*. Lanham, Md.: Rowman and Littlefield.

INDEX

A

Abbey, Edward 5, 183, 185, 199, 207
Acclimatisation Societies 4
Ackerman, Bruce 77, 91, 136
adaptive management 95–6
administrative mind 88
administrative rationalism 75–6
  achievements of 97–8
  crisis of 92–6, 101
  discourse analysis of 86–9
  justification of 89–92
  repertoire of 76–86
Agenda 21 149, 155
agents 18
Aitken, Robert 196
Alakali Inspectorate, United Kingdom 78
Alaska 55, 80–1
Alberta 105
alternative dispute resolution 103
anarchism 207
Anderson, Terry 126, 127, 133, 140, 161n3
Anglers' Cooperative Association 126
animal liberation 186, 214–5
anti-globalization movement 213–4
Australia 4, 104, 150–1
authoritarianism 36–7, 42, 47

B

Babbit, Bruce 119
Bahro, Rudolf 204
Bari, Judi 199
Barnet, Richard 37
Barnett, Harold 52–3, 54
Barry, John 178
Bartlett, Robert 102
basic needs 155
Beck, Ulrich 175–6
Beckerman, Wilfred 53, 54, 57, 60
Becket, Margaret 102
Berejikian, Jeffrey 44
Berger Report 106–7
Berger, Thomas 106–7, 108
Beyond the Limits 33, 35
bioregionalism 188–9
Blair, Tony 151, 178
Blowers, Andrew 175
Blueprint for a Green Economy 130
blueprints for new society 225–6
Booher, David 110
Bookchin, Murray 206–8
Boulding, Kenneth 40
Bradley, Robert L. 55, 67, 68
Braithwaite, John 110
Brechin, Stephen 155
Brent Spar 223
Britain *see* United Kingdom
Brower, David 199–200
Brown, Lester 32, 33, 34, 37, 39
Brundtland, Gro Harlem 16, 122, 145–9, 153–4, 165, 175
Buddhist economics 191
Bureau of Reclamation, United States 126
bureaucracy 89–91
Burford, Anne Gorsuch 62–3
Bush, George H.W. 64, 122, 164
Bush, George W. 7, 46, 62, 65, 66, 78, 82, 84, 122, 152, 164, 205
business, privileged position of 117–18

C

Caldwell, Lynton 81
California 174
capitalism 170–1, 178, 202, 208–9, 225–6, 232
carbon tax 131–2
carrying capacity 27–8, 38, 58
Carter, Jimmy 33, 42
Chain, David 199
Cheney, Dick 65
China 42, 43
Chipko movement 187, 212
chlorofluorocarbons 43–5
Christoff, Peter 173, 174, 189
citizens 113–4
Citizens' Clearinghouse on Hazardous Wastes 210

citizenship 135
  ecological 139, 189
civil society 156, 159
Clayoquot Sound 107
Clean Air Act, United Kingdom 78
Clean Air Act, United States 91, 129, 136
Clear Skies Initiative 66, 84
climate change 45–6, 67–8
Clinton, William 7, 82, 84, 119, 122, 132, 224
Club of Rome 15, 25, 30, 42, 170
Coase, Ronald 127
Cohen, Bernard 59
Coleridge, Samuel 192
comedy 200
command and control 135
Commoner, Barry 200
communication 235
competition 58, 134
complexity 8–9, 69, 90, 93–4, 100, 140, 200–1
compliance 95
computer models 31, 41
congestion charge, London 131
Conservation Movement, United States 14, 76–7
consumerism, green 132, 168, 190, 198
cornucopia 51, 67–9
corporatism 117,165–8 , 177
cost-benefit analysis 83–5
Council of Environmental Experts, Germany 82
Council on Environmental Quality, United States 82
counter-culture, United States 101
courts 79–81
critical realism 23n2
Cronon, William 12
cultural change 198
Czech, Brian 37, 40

D

d'Eaubonne, Françoise 186
Dafis, Cynog 219
Daly, Herman 34–5, 37, 160
Danish Committee on Scientific Dishonesty 56
Darwin, Charles 194
Davidson, Carlos 41
Davis, Karen 196
deep ecology 183–5, 207–8
deliberation 112
democracy 37–8, 99, 108, 147, 158
  ecofeminist 187
  ecological 234–6
  liberal 110–12
democratic pragmatism 99–100, 233
  achievements of 116–7
  as a way of governing 108–113
  discourse analysis of 113–16
  examples of 101–7
  limits of 116–20
Department of Defense, United States 46
Department of the Interior, United States 62–3
Devall, Bill 183, 184
Dewey, John 99
Die Grünen 203–5, 219–21
discourse analysis 11–12, 17–19, 22
discourses
  classifying 13–16
  definition of 9
  environmental 10, 13–14
  history of 19–20
  impact of 20–1
discursive designs 233
Disney, Walt 113, 139
displacement 69, 94
Dobson, Andrew 11
Dobson, Andrew 189, 193, 197
Douglas, Roger 121
Drahos, Peter 110
Du Pont corporation 44

E

Earth Day 49
Earth First! 183, 185, 198–9
Earth in the Balance 177
Earth Liberation Front 4, 199
Earth Summit *see* United Nations Conference on Environment and Development
Easterbrook, Greg 55, 60
Eckersley, Robyn 11, 184
ecocentrism 184
ecofeminism 48, 185–7
ecological economics 34–5
ecological footprint 28, 177
ecological mandarins 37

ecological modernization 143–4, 219–20, 232
as social science theory 180n5
discourse analysis of 169–72
idea of 167–9
reflexive 174, 233–4
strong 173–5, 177, 179, 233
techno-corporatist 172–5
weak 173–5
Ecologically Sustainable Development process, Australia 104, 150–1
eco-Marxism 208–9
economic growth 28, 52, 61, 153, 168
economic rationalism 121–2
assessment of 137–41
discourse analysis of 133–7
economists 52–4
eco-socialism 209
ecosystem management 91–2
ecosystems 34, 38, 111, 133
eco-theology *see* spirituality
Ehrlich, Anne 40, 50n3
Ehrlich, Paul 30, 33, 35, 40, 41, 50n3, 53, 56
Emerson, Ralph Waldo 192
Endangered Species Act, United States 7
energy policy, United States 65
Enlightenment 192–3, 205
entropy 34
environment, concept of 5–6, 139–40
environmental education 200
environmental impact assessment 80–1, 101–2
environmental justice 176, 210–12, 223–4
environmental modernisation 178
environmental pragmatism 100
Environmental Protection Agency, United States 62–4, 78–9, 91, 94, 128–9, 224
environmental security 28, 46
environmentalism of the poor 212–3, 224
environmentality 22
epidemiology 85–6 popular 223
equality, biocentric 184, 194–5
European Union 97, 150
expert advisory committees 82
expertise 87
ecological 37
exponential growth 31–2

F

Factor Four 170
Finland 158, 166
Fischer, Joschka 204, 219
fisheries 129
Foreman, Dave 185, 192, 196, 199, 207
forest certification 109–10
Forest Service, United States 76–7
forests 7–8
Foucault, Michel 22
Fox, Warwick 184
Francis, St. 190
free market environmentalism 122
Friends of the Earth 165
Fundis 204–5

G

Gaia 17, 195–6
Gates, William 28
Georgescu-Roegen, Nicholas 34
German Greens see *Die Grünen*
Germany 131, 164, 167, 177, 203–5, 219–23, 236
Gibbs, Lois 210, 211
Gingrich, Newt 65
Global 2000 Report to the President 33, 42–3, 64
Global Climate Coalition 13
Global Tomorrow Coalition 42
GM Nation, United Kingdom 102
González, George 174, 177
Goodin, Robert 140
Gore, Albert 109, 130, 151, 177–8, 205
governance 96–8, 108–110
green consciousness 182
assessment of 200–2
discourse analysis of 193–7
impact of 197–200
varieties of 183–91
green parties 166, 203–6, 218–21
Germany 203–5, 219–21
United Kingdom 215
United States 205–6
green politics 203
discourse analysis of 215–18
in practice 218–24
varieties of 203–15
green radicalism 16, 179, 181, 234
green taxes 129–32, 138, 140
Greenpeace 7, 223

Guha, Ramachandra 212
Gundersen, Adolf 112, 115

H

Haas, Peter 43
Hahn, Robert 131
Hajer, Maarten 11, 172, 174, 180n
Hardin, Garrett 27, 29, 34, 36, 37, 39, 40, 47, 48, 49, 159, 160
Hassler, William 77, 91, 136
Hawke, Bob 104
Hawken, Paul 172, 178
Hay, Peter 193
Hayek, F.A. von 93
Hays, Samuel 98n1
hazardous wastes 105, 210–11
Healthy Forests Initiative 7, 66
Heilbroner, Robert 36, 39, 191
Heritage Foundation 64
hierarchy 39, 87, 90, 93, 96, 134, 206–7, 216
Hill, Alan 64
Holliday, Charles 152
homo economicus 133–4, 139, 141
horror stories 136
household metaphor 171–2
Huber, Joseph 167

I

idealism 194
immigration 48
implementation deficit 92–3, 96
indigenous peoples 213
industrialism 13–14, 21
Innes, Judith 110
Inspectorate of Pollution, United Kingdom 94–5
institutions 20
insurance companies 46
integrated environmental management 94
integrated pollution control 165–6
International Conference on Population and Development 47
international treaties 64
International Union for the Conservation of Nature 148

J

Jackson, Michael 28
Jacobs, Michael 37, 178
Jahn, Detlef 167
James, William 99
Jänicke, Martin 109
Jänicke, Martin 167
Japan 117, 150, 165, 177
Jevons, William 54
justice 143, 172

K

Kahn, Herman 64
Kelly, Petra 204
Kelman, Steven 112, 115, 120n4, 140
Kemp, Ray 106
Kovel, Joel 208–9
Kyoto Protocol 45–6, 65–7

L

Laduke, Winona 205
Lafferty, William 160
lands, public 124–5
Langhelle, Oluf 147, 175
Lavelle, Rita 63
lay citizen deliberation 105–6
Leal, Donald 126, 127, 133, 140, 161n3
Leavitt, Mike 96, 122
Lee, Kai 91–2, 103
legal corporatism, Germany 87
legitimation crisis 101, 235–6
Leurs, Jeff 4
Lewis, Martin 11, 35
liberalism, green 114
life expectancy, trends in 54–5
lifeboats 36
lifestyle greens 189–90
limits discourse *see* survivalism
Limits to Growth, The 25, 30–3, 53, 170
limits, ecological 25, 30–5, 58, 147, 154, 169–60, 168, 170, 193–4, 216
Lindblom, Charles E. 93, 110, 202
Litfin, Karen 11, 12, 44
Livingstone, Ken 122, 131
Lloyd, William Forster 29
Local Agenda 21 105, 153
Lomborg, Bjørn 55, 56, 57, 58, 60, 66, 68
Lopez, Barry 183
Love Canal 210
Lovelock, James 195
Lowy, Michael 208–9

Luke, Timothy 22, 159
Luther, Martin 140

M

Maathai, Wangari 212
Malthus, Thomas 29, 31
Manes, Christopher 202n1
market liberalism 159–60
markets 121–3, 128, 133, 202
Marxism 47 *see also* eco-Marxism
Meacher, Michael 41, 43
Meadows, Donella H. 31, 35, 43, 44, 49, 146–7, 160
mechanistic metaphor 60, 135, 140
mediation 103
Melbourne 81
metaphors 18–19
Mill, John Stuart 31
misanthropy 184
Miss Ann Thropy 184
Mol, Arthur 175
Monkey Wrench Gang, The 5, 185
Montreal Protocol 43–5, 129
Morse, Chandler 52–3, 54
Muir, John 11
multiculturalism 48
Myers, Norman 34, 37, 39, 53

N

Nader, Ralph 205
Naess, Arne 183
National Coal Policy Project, United States 104
National Environmental Policy Act, United States 80–1, 101
National Environmental Policy Plan, Netherlands 163–4, 167, 169
natural capital 156
Natural Capitalism 178
natural relationships 16–17
natural resource management 76
natural resources 57–8
Nature Conservancy 125
nature, concept of 5, 57–8, 133, 170, 194
Nazis 204
negative feedback 111–2
Negative Population Growth 42
Netherlands 163–4
networks 109, 116, 176, 210–11, 214, 235
New Deal, United States , 77
new environmental policy instruments 109
New Zealand 4
NIMBY 210–11
Nixon 235
Nixon, Richard 63, 235
non-governmental organizations 155
Northwest Power Planning Council 91–2
Norton, Gale 65, 66, 122
Norway 150, 165, 236
nuclear power 6–7, 102, 220

O

O'Brien, Mary 86
O'Connor, James 209
Offe, Claus 205
Official Secrets Act, United Kingdom 107
Ogoni people 213
Olson, Mancur 142n2
ontology 16
Ophuls, William 37, 39, 50n4, 191
organic metaphor 156, 196, 217
Organization for Economic Cooperation and Development 122
Organization for Petroleum Exporting Countries 30, 58
ozone issue 43–5

P

Pacific Cascadia 188–9
Paehlke, Robert 108
patriarchy 186
Pearce, David 130–1
Peccei, Aurelio 40, 50n2
Peirce, Charles 99
Pepper, David 175
piecemeal social engineering 111
Pinchot, Gifford 11, 14, 76–7
Plumwood, Val 187, 192, 193
policy analysis 82–6, 97
policy dialogue 104–5
politicization of science 66, 78–9
pollution 54–5, 57, 93, 127–9, 140, 167–8
pollution control agencies 77–9
Popper, Karl 93, 110–11, 120n3
population 4–5, 58, 59–60, 186
population biology 27–8, 35
population explosion 25, 30, 40
Porritt, Jonathan 118, 191

postmodernism 12
power 117
pragmatism 99–100
precautionary principle 80, 164, 178
preferences, consumer and citizen 113, 119, 139
prices 52–4
privatization 123–8
problem-solving 108–11
  environmental 15, 73
  interactive 110–12, 233
progress 156
Promethean discourse 26, 49, 51–2, 232
  assessment of 67–70
  discourse analysis of 57–61
  impact of 61–7
Prometheus 51
property rights
  in air 125–6
  in water 126
  in wildlife 127
  private 123–8, 140
public consultation 101–2
public inquiries 106–7
public interest 88, 114–5, 118–19
public relations 12–13
public sphere, green 222–3, 226, 234, 236

R

rationality 193, 196
  ecological 119, 217–8
  instrumental 195
  political 119–20
Reagan, Ronald 6, 33, 62, 63, 66, 78, 84, 121, 164
Realos 204–5
Reclaim the Streets 221
reflexive modernity 176
regulation 79–80, 110, 138
Reilly, William 122
Reinventing Government task force 109
resource management bureaucracies 76–7
rhetoric 18–19, 135–6
right-to-know legislation 107–8
risk analysis 85–6
risk society 175–6
risks, environmental 68, 210
rolling rule regime 96–7
Roman Catholic church 47
romanticism 191–3, 195, 199–200, 204, 206
Roosevelt, Franklin 77
Roosevelt, Theodore 76
Rosenbaum, Walter 91
Roskill Commission, United Kingdom 84
Royal Commission on Environmental Pollution, United Kingdom 82
Ruckelshaus, William 63
Russell, Julia 197

S

Sabel, Charles 96
Sagebrush Rebels 62
Sagoff, Mark 113, 115, 139
Salinas de Gotari, Carlos 121
scarcity 53–4
Scarcity and Growth 52–3
Scheinberg, Anne 177
Schlosberg, David 211
Schmidheiny, Stephan 151
Schumacher, E.F. 191
Scientific American 56
Scientific Integrity in Policymaking 78–9
Sessions, George 183, 184
Setting Environmental Standards 178
shadow pricing 83–4
Shell corporation 223
Shiva, Vandana 186–7
Sierra Club 48
Simon, Julian 51, 53, 54, 55, 57, 58, 59–60, 64, 66, 68, 70
Singer, Peter 214–15
Skeptical Environmentalist, The 55–6
social ecology 206–8
social learning 232–3
social movements 205, 221–2
social structure 201–2, 216
Soper, Kate 12
spaceship earth 40
spirituality 186, 190–1
stationary state 31, 35
Stavins, Robert 139
Strong, Maurice 157
structural reform 90–1
Superfund 136
survivalism 15, 25
  origins of 27–9
  assessment of 46–50
  discourse analysis of 38–41

elitism of 38–40, 42, 46
impact of 42–6
political philosophy of 35–8
sustainability 16, 143
Sustainability Index 151, 158, 162–3
sustainable development 143, 172, 175
de-centered approach to 158–9, 233
definitions of 145–7
discourse analysis of 153–57
history of concept 145–6 148–52
prospects for 157–61
Sweden 95, 102, 165–6
systems approach 169–70

T

Thatcher, Margaret 130, 226
thermostat metaphor 115
think tanks 127–8
Third World 148, 172, 212–13
This Common Inheritance 131
Thoreau, Henry David 192
Torgerson, Douglas 88, 145, 200, 214, 233
tradeable quotas 128–9, 138
tragedy of the commons 29–30, 36–7, 40, 123
transpersonal ecology 184
Travolta, John 223
trend evidence 52–5, 60, 69–70
Trump, Donald 28
Twenty-Ninth Day, The 32

U

United Kingdom 80, 130–1, 151, 164, 178
United Nations Conference on Environment and Development 149
United Nations Educational, Scientific, and Cultural Organization
United States 151, 163–4, 177–8

V

Vogel, David 97
Voluntary Human Extinction Movement 41, 184

W

Waldegrave, William 80
Wapner, Paul 66
Warren County, North Carolina 210
Watt, James 11, 62–3
Watts, Philip 152
Weale, Albert 95, 164, 169
Weber, Max 89–90
wetlands 3
Weyerhauser Corporation 13, 117–18
whales 4
White, Lynn 190
Whitman, Christine 65
Wildavsky, Aaron 68, 70, 85
wilderness 5, 185, 213
social construction of 12
Windscale/Sellafield 106, 221
Wise Use Movement 65, 119
Wissenburg, Marcel 114
Wordsworth, William 192
World Bank 150
World Business Council for Sustainable Development 147, 151–2, 175
World Commission on Environment and Development 16, 145, 148
World Economic Forum 151, 158, 162–3
World Summit on Sustainable Development 149–50, 152, 160
World Trade Organization 47, 160, 213
Worldwatch Institute 33, 42, 56

Y

Yellowstone 133